ATP 3-37.10
MCRP 3-40D.13

BASE CAMPS

JANUARY 2017

DISTRIBUTION: Approved for public release; distribution is unlimited.
This publication supersedes ATP 3-37.10/MCRP 3-17.7N, 26 April 2013.

HEADQUARTERS, DEPARTMENT OF THE ARMY

Foreword

This publication has been prepared under our direction for use by our respective commands and other commands as appropriate.

JAMES H. RAYMER
Brigadier General, USA
Commandant
U.S. Army Engineer School

ROBERT S. WALSH
Lieutenant General, USMC
Deputy Commandant for
Combat Development and Integration

This publication is available at the Army Publishing Directorate site (http://www.apd.army.mil), and the Central Army Registry site (https://atiam.train.army.mil/catalog/dashboard) and the USMC doctrine Web site at https://www.doctrine.usmc.mil>.

ATP 3-37.10/ MCRP 3-40D.13

Army Techniques Publication
No. 3-37.10

Headquarters
Department of the Army
Washington, DC

Marine Corps Reference Publication
No. 3-40D.13

Headquarters
Marine Corps Combat Development Command
Headquarters, U.S. Marine Corps
Washington, DC, 27 January 2017

Base Camps

Contents

		Page
	PREFACE	iv
	INTRODUCTION	v
Chapter 1	BASE CAMP OVERVIEW	1-1
	Operational Challenges	1-1
	Basic Considerations	1-3
	Base Camp Life Cycle	1-8
	Base Camp Principles	1-10
	Base Camp Activities	1-12
	Roles and Responsibilities	1-14
Chapter 2	BASE CAMP PLANNING AND DESIGN	2-1
	Planning Principles	2-1
	Development Planning	2-9
	Funding Sources and Authority	2-12
	Base Camp Planning and Design Considerations	2-12
Chapter 3	BASE CAMP CONSTRUCTION	3-1
	Facilities and Infrastructure	3-1
	Contingency Construction	3-2
	General Construction Requirements	3-3
	Construction Methods	3-9
	Construction Means	3-10
	Horizontal-Construction Projects	3-12
	Vertical-Construction Projects	3-13
Chapter 4	BASE CAMP OPERATIONS AND MAINTENANCE	4-1
	Base Camp Operations	4-1

Distribution: Approved for public release; distribution is unlimited.

*This publication supersedes ATP 3-37.10/MCRP 3.17.7N, 26 April 2013.

Contents

	Base Operations Center	4-6
Chapter 5	**BASE CAMP SECURITY AND DEFENSE**	**5-1**
	Base Camp Protection	5-1
	Base Camp Threats	5-2
	Base Camp Protection Framework	5-3
	Base Camp Protection Forces	5-6
	Base Camp Protection Considerations	5-9
	Base Camp Defense Tasks	5-9
	Security Operations	5-10
	Integration of Base Camp Protection	5-18
Chapter 6	**BASE CAMP TRANSFER AND CLOSURE**	**6-1**
	Transfer and Closure Plan	6-1
	General Requirements	6-2
Appendix A	METRIC CONVERSION CHART	A-1
Appendix B	BASE CAMP MASTER PLANNING AND THE MILITARY DECISION-MAKING PROCESS/MARINE CORPS PLANNING PROCESS	B-1
Appendix C	SAMPLE OF ARMY BASE CAMP APPENDIX/ANNEX	C-1
Appendix D	BASE CAMP LAND USE PLANNING	D-1
Appendix E	BASE CAMP PLANNING FACTORS	E-1
Appendix F	BASE CAMP FACILITIES AND INFRASTRUCTURE DESIGN	F-1
Appendix G	REACHBACK	G-1
Appendix H	BASE CAMP COMMUNICATIONS SUPPORT	H-1
	GLOSSARY	Glossary-1
	REFERENCES	References-1
	INDEX	Index-1

Figures

Figure 1-1. Base camp levels of service standards ... 1-6
Figure 1-2. Base camp life cycle ... 1-8
Figure 2-1. Base camp master plan products ... 2-8
Figure 2-2. Base camp development planning process ... 2-9
Figure 3-1. Construction project phasing model ... 3-4
Figure 4-1. Example of a typical BOC organization ... 4-6
Figure 5-1. Framework for base camp security and defense ... 5-4
Figure 5-2. Sidewall and overhead protection ... 5-17
Figure 5-3. Tower location and unobservable areas ... 5-18
Figure C-1. Sample Army base camp appendix/annex ... C-2
Figure D-1. Sample land use plan ... D-2
Figure D-2. Rectangular box design ... D-3
Figure D-3. Wheel design ... D-4

Figure F-1. Base camp power life cycle F-10
Figure G-1. UROC Web site G-2
Figure G-2. REDi home page G-3
Figure H-1. Example of Army signal support configuration for a company base camp H-4
Figure H-2. Example of Army signal support configuration for a battalion/battalion landing team base camp H-6
Figure H-3. Example of Army signal support configuration for a support area base camp H-7
Figure H-4. MAGTF communication architecture overview H-8

Tables

Introductory table 1. Modified Army/Marine Corps terms vi
Table 1-1. Threat levels and capabilities required 1-3
Table 1-2. Base camp duration 1-4
Table 1-3. Base camp sizes and approximate populations 1-5
Table 2-1. General design considerations in relation to base camp master planning principles 2-14
Table 5-1. Summary of base camp protection forces 5-7
Table A-1. Metric conversion chart A-1
Table B-1. Base camp planning considerations during the planning process B-2
Table B-2. Site selection considerations in relation to mission variables (METT-TC/METT-T) B-5
Table B-3. Terrain considerations in relation to OAKOC/KOCOA B-8
Table D-1. Typical land use categories for base camps D-5
Table E-1. Base camp sizes and planning factors E-2
Table E-2. Sample contingency standards E-6
Table E-3. Sample planning factors for personnel accommodations for temporary standards E-7
Table E-4. Sample maximum numbers of personnel structures E-8
Table E-5. Expansion increase planning factors E-9
Table E-6. Planning factors for unit headquarters at a BCT/RCT size base camp E-9
Table E-7. Example of planning factors for office space E-10
Table E-8. Maximum average estimated cost for general purpose medium base camp E-10
Table E-9. Maximum average estimated cost for SEA huts/SWAhuts base camp E-11
Table E-10. Sample planning factors for medical treatment facilities E-11
Table E-11. Grossing factors E-11
Table E-12. Base camp utilities planning factors E-12
Table E-13. Power generation options versus costs E-12
Table E-14. Power system planning considerations E-12

Preface

ATP 3-37.10/MCRP 3-40D.13 is a compilation of tactics, techniques, and procedures (TTP) found in doctrine, lessons learned, and other reference material that provides an integrated, systematic approach to base camps. It codifies the recent efforts of the Base Camp Integrated Capabilities Development Team as part of the Army capabilities-based assessment process and serves commanders and their staffs as a comprehensive how-to guide for performing all activities of the base camp life cycle during deployments.

This manual acknowledges that each base camp will be unique, based on mission requirements and the theater-specific facility allowances and construction standards that apply. Therefore, this manual relies on its user's ability to apply experience and good judgment in incorporating the base camp principles and procedures that are provided here, along with the wisdom to seek out the necessary expertise where needed in generating options and implementing best practices that result in efficient and effective base camps.

The principal audience for ATP 3-37.10/MCRP 3-40D.13 is all members of the profession of arms. Commanders and staffs of a joint task force or multinational headquarters should also refer to applicable joint or multinational doctrine concerning the range of military operations and joint or multinational forces. Trainers and educators throughout the Army and Marine Corps will also use this manual.

Commanders, staffs, and subordinates ensure that their decisions and actions comply with applicable United States (U.S.) international, and in some cases, host-nation (HN) laws and regulations. Commanders at all levels ensure that their Soldiers/Marines operate in accordance with the law of war and the rules of engagement. See FM 27-10 for more information.

ATP 3-37.10/MCRP 3-40D.13 uses joint terms where applicable. Selected joint and Army/Marine Corps terms and definitions appear in the glossary and the text. Terms and definitions for which ATP 3-37.10/MCRP 3-40D.10 is the proponent publication (the authority) are indicated in the glossary and are printed in boldface and italicized in the text. These terms and their definitions will be incorporated into the next revision of ADRP 1-02 and MCRP 1-10.2. For other definitions in the text, the term is italicized, and the number of the proponent manual follows the definition.

ATP 3-37.10/MCRP 3-40D.13 applies to the Active Army, Army National Guard/Army National Guard of the United States, Marine Corps, United States Army Reserve, and Marine Corps Reserve unless otherwise stated.

The proponent of ATP 3-37.10/MCRP 3-40D.13 is the Maneuver Support Center of Excellence (MSCoE). The preparing agency is the MSCoE Capabilities Development and Integration Directorate (CDID); Concepts, Organizations, and Doctrine Development Division (CODDD); Doctrine Branch. Send comments and recommendations on DA Form 2028, *Recommended Changes to Publications and Blank Forms*, to Commander, MSCoE, ATTN: ATZT-CDC, 14000 MSCoE Loop, Suite 270, Fort Leonard Wood, MO 65473-8929; by e-mail to usarmy.leonardwood.mscoe.mbx.cdidcodddengdoc@mail.mil; or submit an electronic DA Form 2028. The United States Marine Corps proponent for this publication is the United States Marine Corps Engineer School. Submit changes to United States Marine Corps Engineer School, MAGTF Engineer Center, BB-12, RM 245, Camp Lejeune, NC 28542-0069.

Introduction

Meeting America's strategic objectives hinges on the ability to rapidly deploy forces at any time, in any environment, and against any adversary. Contingency basing provides strategic integration of policy, planning, and resourcing for enduring and semipermanent deployment locations. See DODD 3000.10 for more information about contingency basing integration from the joint perspective. Bases and base camps are physical locations designated in contingency plans to sustain and protect deploying forces. This publication covers up to the semipermanent construction duration.

Exploiting this expeditionary capability often places units in an austere operational environment that is inherently uncertain, with poor or war-affected infrastructure that cannot accommodate deployed forces. Establishing base camps helps extend and maintain operational reach and are vital in projecting and sustaining combat power. Creating base camps that are efficient and effective helps conserve resources, protects and sustains forces, limits liabilities, and reduces the overall sustainment/logistics burden during extended operations—ultimately enabling mission success.

Operating from base camps is a fundamental tactic of Armed Forces. Recent experiences in contingency operations overseas have revealed some of the challenges that base camps present to commanders. These experiences have also exposed the consequences when the activities of the base camp life cycle are inadequately considered or addressed during the course of operations. Some of these consequences include—

- The inefficient use of resources such as time, materials, water, energy, fuel, and money.
- Hazards (such as fire and electrocution) associated with improper construction.
- Health-related concerns associated with trash burning and improper waste management.
- Negative effects on time and money for base camp transfers and closures due to residual environmental issues.

Although contingency operations are generally thought of as short in duration, many situations in the past have resulted in forces remaining in operational areas far longer than anticipated. Often, bivouac sites, assembly areas, and existing facilities occupied during the course of operations become de facto base camps. These impromptu facilities and infrastructure evolve without the necessary prerequisite planning and incorporation of appropriate design, construction, and protection considerations. These ad hoc base camps typically exhibit flaws and yield inefficiencies that waste limited resources, pose hazards to occupants and the environment, and ultimately detract from the overall mission.

Base camps in support of operations range from platoon to support area size camps, with varying levels of capabilities and construction standards that are indicative of the anticipated life span. Base camps routinely support U.S. and multinational forces and other unified action partners, operating anywhere along the range of military operations.

Base camps may be decisive points within lines of operations for achieving mission objectives in a majority of contingencies. As part of the contingency basing strategy for the operational area, base camps must be viewed through a life cycle construct that includes the development of base camps from pre-establishment through transfer or closure, with levels of increasing base camp capabilities. The driving forces throughout the base camp life cycle are objectives that emanate from the top level commander, who drives policy and ensures strategic synchronization, to the operational commander who owns the area of operations (AO), to the base camp commander/base operating support–integrator (BOS-I) and the commanders of tenant units.

Enduring and contingency are words that nest with duration characterizations described in JP 3-34. For the purposes of this book, those words are depicted only to nest duration times with the vertical integration of JP 3-34. The words enduring and contingency have been commonly associated with base camps; however, they do not enhance meaning and often only contribute to confusion. One of the developments in this manual is the base camp classification system. This provides a simple way to classify base camps and mitigate the confusion that has been generated by the inconsistent application of various naming conventions, such as contingency operating base and main operating base. Regardless of how base camps may be referred to by

Introduction

Service components or their commands, all base camps are broadly classified by their size, level of services, and purpose.

This manual uses the term planning process to indicate the military decision-making process (MDMP)/Marine Corps Planning Process (MCPP) and troop-leading procedures. Battalion size and larger units use the MDMP or the MCPP, depending on their Service. Company size and smaller units follow troop-leading procedures.

This manual uses the term mission variables to indicate the Army and Marine Corps uses of the term. For the Army, mission variables consist of mission, enemy, terrain and weather, troops and support available– time available and civil considerations (METT-TC). For the Marine Corps (and in joint doctrine) mission variables consist of mission, enemy, terrain and weather, troops and support available–time available (METT-T).

When this manual uses two terms separated by a slash (/), the first term is the Army term; and the second term is the Marine Corps term. Key differences in Army and Marine Corps terms include—

- (Army) brigade combat team (BCT)/(Marine Corps) regimental combat team (RCT) (written in this manual as BCT/RCT).
- (Army) decisive action/(Marine Corps) simultaneous activities (written in this manual as decisive action/simultaneous activities).
- (Army) geospatial engineer/(Marine Corps) geographic intelligence specialist (written in this manual as geospatial engineer/geographic intelligence specialist).
- (Army) intelligence preparation of the battlefield (IPB)/(Marine Corps) intelligence preparation of the battlespace (IPB) (written in this manual as IPB).
- (Army) memory aid expressed as observation and fields of fire, avenues of approach, key terrain, obstacles, and cover and concealment (OAKOC)/(Marine Corps) key terrain, observation and fields of fire, cover and concealment, obstacles, and avenues of approach (KOCOA) (written in this manual as OAKOC/KOCOA).
- (Army) movement and maneuver warfighting function/(Marine Corps) maneuver warfighting function (written in this manual as movement and maneuver/maneuver).
- (Army) protection warfighting function/(Marine Corps) force protection warfighting function (written in this manual as protection/force protection).
- (Army) running estimate/(Marine Corps) staff estimate (written in this manual as running estimate/staff estimate).
- (Army) situational understanding (SU)/(Marine Corps) situational awareness (SA) (written in this manual as SU/SA).
- (Army) standard operating procedure (SOP)/(Marine Corps) (SOP) (written in this manual as SOP).
- (Army) sustainment/(Marine Corps) combat service support.
- (Army) sustainment warfighting function/(Marine Corps) logistics warfighting function (written in this manual as sustainment/logistics).
- (Army) signal/(Marine Corps) communications (written in this manual as signal/communications).
- (Army) unified action partners/(Marine Corps) interorganizational partners (written in this manual as unified action/interorganizational partners).

The development of this manual resulted in the modification of Army/Marine Corps terms (see introductory table 1).

Introductory table 1. Modified Army/Marine Corps terms

Term	Remarks
base camp	New Army/Marine Corps definition. (ATP 3-37.10/MCRP 3-40D.13 is now the proponent manual.)

Introduction

ATP 3-37.10/MCRP 3-40D.13 covers the following information:
- **Chapter 1** provides an overview of base camps and describes some of the challenges in establishing and maintaining them in future operational environments. It describes the base camp life cycle and the inherent roles and responsibilities and offers principles that planners and executors incorporate to optimize efficiency and achieve effectiveness.
- **Chapter 2** discusses strategic, operational, and tactical planning with a focus on how commanders and their supporting staffs at the operational and tactical levels use MDMP/MCPP to determine their requirements for base camps and integrate base camps into the concept of operations. It also provides an overview of the base camp development planning process that is performed once the decision is made to establish a base camp.
- **Chapter 3** focuses on the base camp design and construction and the balancing of tactical, operational, sustainment, and engineering requirements for designing facilities and infrastructure to fulfill the purpose of the base camp and its functional requirements based on needs.
- **Chapter 4** focuses on base camp operations and maintenance and the means, methods, and procedures for fulfilling construction requirements.
- **Chapter 5** provides information on conducting base camp security and defense measures as part of the overall protection plan for base camps.
- **Chapter 6** describes the organizational structuring required for operating and managing base camps and centers on the operation of base camp management centers and base operations centers (BOCs). It also discusses three critical functional areas for base camps—emergency management, master planning, and contract management.
- **Appendix A** contains a metric conversion chart.
- **Appendix B** discusses base camp master planning and the military decision making process.

For Marine Corps users: Appendix B references Army annexes, appendixes, and tabs that do not align with Marine Corps annexes, appendixes and tabs. See MCWP 5-10 for correct Marine Corps annexes, appendixes, and tabs.

- **Appendix C** provides a sample base camp appendix/annex.

For Marine Corps users: Appendix C references Army annexes, appendixes, and tabs that do not align with Marine Corps annexes, appendixes and tabs. See MCWP 5-10 for correct Marine Corps annexes, appendixes, and tabs.

- **Appendix D** describes mapping land use development (base camp layout), depicting available land (lease boundaries, explosive-hazard areas, and environmental surveyed areas), and supporting construction project lists.
- **Appendix E** discusses base camp planning factor estimates.
- **Appendix F** provides facilities and infrastructure design information for effective base camp systems.
- **Appendix G** provides points of contact for reachback capabilities for those specialty areas not typically organic to the base camp command/BOS-I and staff.
- **Appendix H** describes communications support requirements for base camps and the roles and responsibilities for requirements.

A complete listing of preferred metric units for general use is contained in Federal Standard 376B http://www.nist.gov/pml/wmd/metric/upload/fs376-b.pdf.

United States Marine Corps publication numbers have been updated on https://www.doctrine.usmc.mil to a revised numbering system. Not all downloadable publications depict the new numbers on the source document.

This page intentionally left blank.

Chapter 1
Base Camp Overview

Force projection is the ability to project the military instrument of national power from the United States or another theater, in response to requirements for military operations. (JP 3-0.) The capability to deploy forces and rapidly integrate them into an operational area is essential. Consequently, base camps have been constructed to support ground forces in virtually every contingency operation and exist to protect forward deployed forces while they prepare to execute tactical operations in support of the mission. The long-term U.S. national military strategy anticipates extended deployments in austere parts of the world having limited infrastructure. Since living quarters, dining facilities, recreation facilities, and other support facilities are all important components of a military mission, extended deployments require base camps. This chapter provides an overview of base camps and includes discussion on classifying base camps based on size, construction standard, level of services, and purpose. It includes some of the challenges that base camp commanders/BOS-Is will face in establishing base camps in today's operational environment. It also provides principles that are incorporated throughout the life cycle and the roles and responsibilities of base camp commanders, staff, and BOS-Is. See JP 4-0 for more information on BOS-I.

OPERATIONAL CHALLENGES

1-1. Developing base camps is a complex task that balances mission, protection, sustainment, and construction requirements. This task is further complicated by changes in missions, fluctuating troop levels, turbulent civil conditions, threat factors, evolving end states that are inherent in contingency operations, and the fact that base camps routinely support unified action/interorganizational partners operating anywhere along the range of military operations. Added to this are time and resource constraints, funding and contract restrictions, theater entry conditions, mission duration, access to resources, competing requirements, and environmental considerations. See EP 1105-3-1 for more information.

1-2. Establishing base camps is resource-intensive—not only in terms of the labor, equipment, and materials needed for the construction and operation and maintenance (O&M) of facilities and infrastructure, but also the command and staff efforts that are required throughout the base camp life cycle. Limited base camp planning, designing, and managing assets within the operational force, which are essentially nonexistent at the lower tactical levels, further exacerbate the problem. This demand on commanders and staffs is generated for all base camps, regardless of size, function, or military mission.

1-3. Often, the only differences in the effort required between a platoon and a battalion base camp are the amount and type of resources expended and the degree of technical and engineering expertise required. Commanders, supported by their staffs, primarily overcome these challenges by anticipating and identifying base camp requirements and any shortfalls in capabilities for each phase of the operation as early as possible during planning. This facilitates the timely augmentation of the necessary base camp capabilities through force tailoring and task organization.

1-4. Commanders and staffs are challenged to achieve the level of desired responsiveness in establishing base camps to effectively fulfill mission requirements. Programming and funding procedures for base camp development, especially in deployed locations, are cumbersome and often cannot keep pace with rapidly changing mission requirements. Identifying base camp requirements as early as possible during planning provides lead time that will help ensure that base camp requirements are fulfilled in a timely manner. As with all missions, the purpose of each base camp must be clearly stated.

ENTRY CONDITIONS

1-5. The level of support or hostility that U.S. forces encounter has a significant effect on developing base camps. The deployment of forces may be opposed or unopposed by an enemy. U.S. forces seek an unopposed entry, with or without HN assistance. An assisted entry requires HN cooperation. Consequently, the HN may provide facilities for deploying forces. In an unassisted entry, no secure facilities for deploying forces exist. In an unopposed entry, operations may be conducted from base camps once they are established.

1-6. U.S. forces operating with the assistance of the HN government can reasonably assume some level of support from the local population. This situation eases base camp development by setting conditions that may offer easier and more reliable access to resources as well as assistance from the local population in obtaining construction materials and contracted labor. It also facilitates the early reconnaissance of potential base camp locations, which enables planning and design.

1-7. In situations where U.S. forces are assisted by the HN does not infer that all entities in the government or in local population will be supportive. These situations may impede access to resources and contract labor, require greater security measures, and limit the ability of U.S. forces to conduct an early reconnaissance of potential base camp locations. These locations may also be driven, at least initially, by tactical rather than sustainment considerations. A similar situation might involve humanitarian assistance and foreign disaster relief tasks in response to natural or man-made disasters, where the damage caused by the disaster produces many of the same effects: limited access to resources, complicated transportation, and limited initial reconnaissance.

1-8. Opposed entry requires that U.S. forces conduct forcible-entry operations to gain a foothold or lodgment in a foreign country. These situations are very difficult to plan for, as access to potential base camp locations is limited. Some base camps may be designated in the lodgment area to support entry, while others may not be established until commanders decide when, where, and for what purpose to conduct operations from base camps. A hostile government or population limits access to resources (quantity and types available), and base camp location selection must incorporate the tactical situation and other considerations of the operational environment.

THREATS

1-9. There are three levels of threats against bases and base camps. For each level, there is a general description and categorization of threat activities, recommended security responses to counter them, and establish a common reference for planning. Table 1-1 lists the threat levels and the capabilities required. Each level or any combination of levels may exist in the AO independently or simultaneously in the base camp. Emphasis on specific base security measures may depend on the anticipated level of threat. This does not suggest that threat activities will occur in a specific sequence or that there is a necessary interrelationship between each level. The base camp commander/BOS-I and staff should go beyond the size and type of units when determining and describing levels of threat.

1-10. Threat levels should be based on the activity, capability, and intent of enemy agents or forces. They can be further described by looking at mission effect. Whereas a Level II threat requires a response by a mobile security force (MSF) and may have a measurable effect on the mission, a Level III threat could cause mission failure and requires a significant tactical combat force response.

Table 1-1. Threat levels and capabilities required

Threat Level	Examples	Capability Requirements
I	Agents, saboteurs, sympathizers, terrorists, insider threats, and civil disturbances	Organic base security forces.
II	Small tactical units; irregular forces may include significant stand-off weapons threats	Base security forces plus a mobile security force.
III	Large tactical force operations, including airborne, heliborne, amphibious, infiltration, and major air operations	Base security forces, a mobile security force, and a tactical combat force.

UNCERTAINTY

1-11. Planning and designing scalable base camps help to mitigate the effects of uncertainty. Commanders and their staffs must be tolerant of the uncertainties associated with establishing base camps in support of contingency operations and be prepared to handle the inherent ambiguities and complexities through extensive planning and continuous coordination that effectively mitigate risk. Two of the most demanding challenges are accurately estimating the intended base camp population (personnel, vehicles, and equipment on the base camp at any one time) and determining the expected life span of the base camp based on mission duration. The size and composition of the deployed force may change between planning and construction and will almost certainly change over the life span of a base camp. These uncertainties force planners to plan and design base camps based on valid assumptions—which if proven false, can result in inadequate facilities and infrastructure or wasted construction.

BASIC CONSIDERATIONS

1-12. A *base camp* is an evolving military facility that supports the military operations of a deployed unit and provides the necessary support and services for sustained operations. Army and Marine Corps basing typically fall into two general categories: permanent (bases or installations) and nonpermanent (base camps). Bases or installations consist of permanent facilities and are generally established in HNs where the United States has a long-term lease agreement and a status-of-forces agreement. Base camps are nonpermanent by design and are designated as a base only when the intention is to make them permanent. Base camps may have a specific purpose, or they may be multifunctional. While base camps are not permanent bases or installations, the longer they exist, the more they exhibit many of the same characteristics in terms of the support and services that are provided and types of facilities that are developed.

1-13. Base camps provide a protected location from which to project and sustain combat power. Commanders apply operational art to decide when, where, and for what purpose to operate from base camps. Strategic and operational reach may initially depend on existing bases/base camps. Extending that reach and prolonging endurance to achieve success will likely require the forward positioning of base camps along lines of operations. The arrangement and location of base camps (often in austere, rapidly-emplaced configurations) complement the ability of land-based forces to conduct sustained, continuous operations to operational depth by providing locations throughout the operational area to sustain and project combat power.

1-14. A base or base camp can contain one or more units from one or more Services and typically support U.S. and multinational forces and other unified action/interorganizational partners operating anywhere along the range of military operations. A base or base camp has a defined perimeter and established access controls and should take advantage of natural and man-made features.

1-15. A commander designates an area or facility as a base or base camp and often designates a single commander as the base or base camp commander/BOS-I responsible for protection, terrain management, and day-to-day operations of the base or base camp. This allows other units to focus on their primary functions. Units located within the base or base camp are under the tactical control of the base or base camp commander/BOS-I for protection and sustainment. Within large echelon and support areas, controlling commanders may designate base clusters for the mutual protection and accomplishment of mission objectives.

1-16. Base camps may be used for an extended time and are often critical to wide area security. See ADRP 3-0. During protracted operations, they may be expanded and improved to establish a more permanent presence. The scale and complexity of a base camp, however, are generally related to the size and nature of the force that it supports. The decision to expand or improve a base camp must support the basing strategy, the commander's intent, and the concept of operations.

CLASSIFICATION

1-17. Base camps are broadly classified by duration, purpose, and size. This classification system provides common terminology and a framework that aids in the conduct of all base camp life cycle activities.

Base Camp Duration

1-18. A base camp may be classified according to its expected duration as shown in table 1-2. A contingency base camp is expected to operate 2–10 years or less, while an enduring base camp is expected to operate more than 5 years or longer. Facilities should transition from contingency to enduring standards when appropriate, typically any time within a 6-month to 5-year period. These timelines provide a framework to plan for the transition of standards. The actual trigger for transition is based on conditions and other factors.

Table 1-2. Base camp duration

Phase	Construction Standard	Expected Duration
Contingency	Organic	Up to 90 days
	Initial	Up to 6 months
	Temporary	Up to 5 years
	Semipermanent	2-10 years
Enduring	Permanent	5 years or longer

1-19. Expected base camp duration affects the construction standards used for facilities and infrastructure. While enduring construction standards are not typically used during the contingency phase of an operation, at times semipermanent construction standards may sometimes be used in place of initial (completed with organic equipment) or temporary construction when site considerations require or mission parameters lead to their use. The combatant commander (CCDR), in coordination with Service components and the Services, specifies the construction standards for facilities in the theater to optimize the engineer effort expended on any given facility while assuring that the facilities are adequate for health, safety, and mission accomplishment.

Base Camp Purpose

1-20. Base camps are developed for a specific purpose. A base camp can serve as an intermediate staging base, a forward operating base, or a sustainment/logistics base; support reception, staging, onward movement, integration, training, or detention facilities; or perform multiple functions. The designated purpose and the operational requirements of tenant units serve as the primary guide in designing the base camp.

Base Camp Size

1-21. There are five sizes of base camps: platoon, company, battalion/battalion landing team, BCT/RCT, and support area. Table 1-3 shows base camp sizes and the populations associated with each. The base camp population typically includes tenant and transient units and organizations, including U.S., multinational, and HN personnel, units, and organizations to include contractors authorized to accompany the force (CAAF) and selected non-CAAF. Transient units and organizations are those that come to the base camp for specified services and support. This may not necessarily include staying overnight. Determining the number of transients that a base camp can serve and understanding service and support relationships with other base camps are critical factors in accurately identifying requirements for base camp facilities and infrastructure, services, and support.

Table 1-3. Base camp sizes and approximate populations

Base Camp Size	Approximate Population
Platoon	50
Company	300
Battalion/battalion landing team	1,000
BCT/RCT	3,000
Support area	6,000 or greater
Legend:	
BCT/RCT brigade combat team/regimental combat team	

BASE CAMP STANDARDS

1-22. CCDRs establish theater base camp standards that are tailored for the joint operations area that provides guidance on facility allowances and standards for construction, quality of life (QOL), design, environmental, and force protection issues. Base camp standards are developed using operational and mission variables. Additionally, combatant commands (CCMDs) consider the unique characteristics of the region and the anticipated duration of a mission in their basing standards. For example, if wooden huts are approved for temporary construction [in Southeast Asia (SEA)/Southwest Asia (SWA)], it may be more cost effective—based on resources, local climate and insects, and the local labor market—to use concrete masonry unit (CMU) construction instead. See JP 3-34, UFC 1-201-01 and applicable CCDRs guidance for more information on basic guidelines for facility allowances and construction standards.

CONSTRUCTION STANDARDS

1-23. There are three construction standards for base camps: initial, temporary, and semipermanent. The time periods for each standard are derived from the expected design life—not how long a facility may actually be used. Commanders ensure that subordinate units tasked to perform base camp construction operations have the necessary capabilities, through augmentation as necessary, to execute base camp construction tasks to standard based on a troop-to-task analysis.

Initial

1-24. The initial construction standard is characterized by facilities with minimum capabilities, requiring minimal engineer effort; and it simplifies material transport and availability. It is intended for immediate use by units upon arrival in theater up to 6 months. Typical with transient mission activities, base camps may require system upgrades or replacement by more substantial or durable facilities during the course of operations.

1-25. Organic construction is a subset of the initial construction standard. It is intended for the immediate use by units upon arrival in theater for up to 90 days; however, it may be used for up to 6 months. Units use their organic/table of organization and equipment capabilities to the fullest extent possible to construct base camps. Organic capabilities may vary based on the type of unit, training, experience, and equipment available. They typically provide for initial force presence and maneuver activities until force flow supports the arrival of engineer resources.

Temporary

1-26. Temporary construction is the low-cost construction of buildings and facilities designed to serve a life expectancy of 5 years or less. A minimal facility is intended to increase the efficiency of operations and moderately improve the QOL for occupants; maintainability is a secondary consideration. Temporary construction is characterized by low cost, expedient, construction using locally available materials, construction methods, and equipment. Temporary construction typically cannot be economically converted to a higher construction level. Temporary standard construction can be used from the start of an operation if directed by a CCDR.

Chapter 1

Semipermanent

1-27. Semipermanent construction refers to buildings and facilities designed and constructed to serve a life expectancy of less than 10 years. With maintenance and upkeep of critical building systems, the life expectancy of a facility can be extended to 25 years. The expediency of construction and material availability may be factors. Facilities are intended for a more enduring presence, with operational characteristics and functional performance similar to permanent construction. The types of structures used depends on duration. Temporary construction may be used initially if directed by the CCDR after carefully considering the political situation, cost, QOL, and other criteria.

BASE CAMP LEVELS OF SERVICES STANDARDS

1-28. As shown in figure 1-1, there are three increasing levels of service standards for base camps: basic, expanded, and enhanced. Service standards describe the characteristics of a base camp in terms of support and services (and overall QOL) that are provided and the nature of the construction effort applied, commensurate with the anticipated duration of the mission. There is no direct link between construction standards and QOL standards. For example, a base camp may have initial construction standards with expanded QOL standards. The CCDR sets QOL standards for each level depending on local conditions. Base camps in support of short-duration missions are more austere and require fewer resources to establish and operate, while those for longer-duration missions generally require greater resources. Not all similar-size base camps have the same services, and the implementation of these capabilities is not directly linked to operational phases.

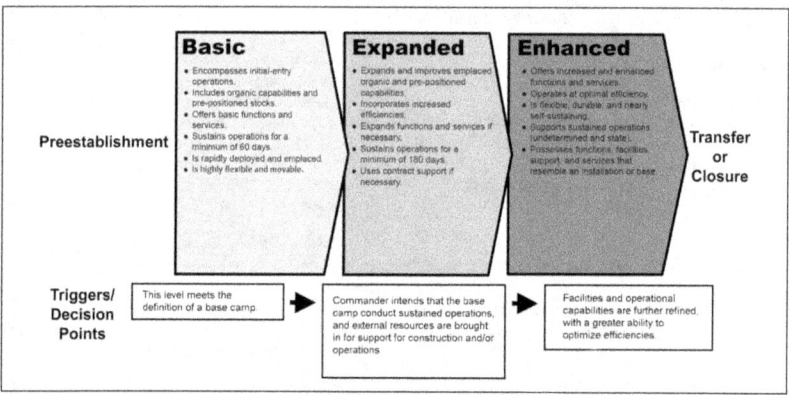

Figure 1-1. Base camp levels of service standards

1-29. Service standards are scaled within the three levels. Scaling refers to the quality of the service provided, not the number or types of service provided.

1-30. Changing the level of service of a base camp should be a deliberate decision that is linked to a decision point in the operational plan or triggered by a clearly identifiable change in the situation. Each base camp has a planned life cycle that can be adapted as the operation progresses. Incorporating the base camp principle of scalability facilitates expansion or reduction without a major redesign of the base camp. When a decision point or trigger to change the base camp services cannot be clearly established based on the uncertainty of the situation, the base camp commander/BOS-I and planners anticipate the requirements necessary to achieve the next capability and include them within the base camp master plan that is linked to the basing strategy. This allows for a timely response once a decision is made to change the level of services of the base camp.

1-31. Units use their organic construction capabilities to the fullest extent possible to construct base camps to the directed standard. Organic construction capability varies based on the type of unit, training, experience, and equipment available. For example, an infantry unit augmented with an engineering capability may be able to construct some facilities to the initial or temporary standard, while a general engineer unit with a greater organic construction capability may be able to construct some facilities to the semipermanent standard. Commanders ensure that subordinate units tasked to perform base camp construction operations have the necessary capabilities, through augmentation as necessary, to execute base camp construction tasks to standard based on a troop-to-task analysis.

Basic

1-32. Basic services are established as part of the initial entry and are primarily implemented using organic capabilities and pre-positioned stock. Basic services are those functions and services that are considered essential for sustaining operations for a minimum of 60 days. Basic services are characterized by rapid deployment and emplacement. Basic facilities and infrastructure are highly flexible and movable and are constructed with minimal engineering effort. Construction of these facilities takes full advantage of unit organic capabilities. The facilities are intended for the immediate operational use by units upon arrival for up to 6 months. Basic facilities will follow initial construction standards.

1-33. Units can provide the basic QOL standard with their organic capabilities or other operational unit capabilities. There is little to no contracted support affiliated with basic QOL standards.

Expanded

1-34. Expanded services are those that have been improved to increase efficiencies in the provision of base camp support and services and expanded to sustain operations for a minimum of 180 days. Expanded facilities are constructed from additional engineer efforts above the basic facility standards. They are intended to increase operational efficiency for use up to 2 years and may be used to fulfill requirements up to 5 years. Because temporary construction is not intended for long-term use, extending the life of these facilities and infrastructure through modifications and increased maintenance and repairs is generally more expensive than building semipermanent facilities and infrastructure from the start. Therefore, commanders should strive to designate enduring base camps as early as possible. Expanded facilities comply with initial or temporary joint construction standards. For example, a prime power system may be installed, a water-bottling plant may replace imported bottled water, or an existing facility may be upgraded to replace tents. Engineer units or contracted support may be used to achieve the desired results. See JP 4-10 for more information on contracted support.

1-35. The expanded QOL standard is based on support and services beyond operational unit capabilities and involves contracted support or specialized military units and organizations. The expanded QOL is intended to decrease the stress on personnel deployed for longer periods of time.

Enhanced

1-36. Enhanced services surpass expanded services. They have been improved to operate at optimal efficiency and sustained operations for an unspecified duration. These services are flexible, durable, nearly self-sustaining, and primarily implemented through contracted support. Many of the functions, facilities, and services and much of the support resemble that of a permanent base or installation. Enhanced facilities allow for finishes, materials, and systems selected for moderate energy efficiency, maintenance, and life cycle cost. These facilities are intended for a life expectancy of more than 2 years to less than 10 years. Enhanced facilities comply with semipermanent or permanent joint construction standards. Department of Defense (DOD) construction agents (United States Army Corps of Engineers [USACE], Naval Facilities Engineering Command [NAVFAC], or other such DOD-approved activity) are the principal organizations to design, award, and manage construction contracts in support of enduring facilities according to the applicable Unified Facilities Criteria (UFC). The enhanced QOL standards approach those of an installation. Normally, these enhanced QOL standards should not exceed those of a permanent base or installation.

BASE CAMP LIFE CYCLE

1-37. The base camp life cycle shown in figure 1-2 embodies the major activities that are involved in base camps. These activities are mutually reinforcing, not mutually exclusive, and include—
- Strategic system and policy integration.
- Planning and design.
- Construction.
- Operations and maintenance.
- Transfer and closure.
- Command and Control

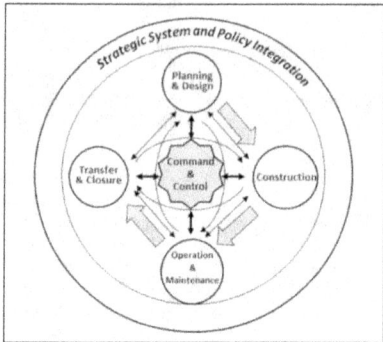

Figure 1-2. Base camp life cycle

1-38. The base camp life cycle includes four activities that relate to the actual life span of a base camp, which are planning and design, construction, operations and maintenance, and transfer or closure. These four activities are usually sequential, although they are recurring and often overlap because base camps are dynamic—continuously modified and improved based on threat; mission requirements; and the need for relocation, expansion, or reduction. The base camp life cycle is not directly linked to operational phases, as base camps can be planned, constructed, expanded, or transferred or closed during any phase of an operation.

STRATEGIC SYSTEM AND POLICY INTEGRATION

1-39. The life cycle is encompassed by strategic system and policy integration, which emphasizes that base camps are a system of systems governed by policies and procedures established at the national and service levels. Base camp efforts are integrated as a holistic system to provide consistent policy and doctrine, comprehensive training, integrated command and staff functions, and coordinated resource support, which enables the other life cycle activities. Efficiencies and effectiveness are gained by DOD efforts across doctrine, organization, training, materiel, leadership and education, personnel, and facilities and common Service standards. The CCDR confirms and defines base camp policy and ensures strategic synchronization through the creation of the basing strategy that is reflected in plans and orders and passed to the operational commander who is responsible for the AO, to the base camp commander/BOS-I, and commanders of tenant units.

PLANNING AND DESIGN

1-40. Planning and design are interdependent. Effective design hinges on the accuracy of the information generated during planning, particularly information related to facility and infrastructure requirements, available resources, construction means, and site location. The failure to remain continuously linked with mission planning as it progresses or the mistake of designing in a vacuum can result in design solutions that are unsustainable based on the concept of operations or inadequate in meeting the needs of the commander.

CONSTRUCTION

1-41. Construction, as part of the life cycle, refers to the means and methods devised through planning and design for constructing, modifying, upgrading, and deconstructing base camp facilities and infrastructure. Construction is performed by military units, CAAF, non-CAAF, or any combination necessary to achieve the desired results. Facilities and infrastructure are built using various methods that are evaluated and determined during planning and design. Existing facilities and infrastructure are used to the fullest extent possible to minimize the overall construction effort and reduce the sustainment/logistics footprint. The use of modular systems and prefabricated or pre-engineered components is maximized to facilitate speedy development, achieve scalability, and reduce the time needed for closing base camps.

OPERATIONS AND MAINTENANCE

1-42. The base camp facilities and infrastructure can involve complex systems requiring specialized skills to operate and maintain them. This includes power, water, environmental, and waste management systems. Base camp O&M focuses on facilities and infrastructure and the provision of services and support that fulfill the designated purpose and functional requirements of the base camp. A BOC is commonly established as a centralized activity for directing and controlling base camp operation to facilitate effective base camp management.

TRANSFER OR CLOSURE

1-43. All or portions of a base camp may be closed when no longer needed or transferred to another Service, multinational force, governmental or nongovernmental organization, or the HN. As the operation progresses and mission objectives are achieved, base camps are often realigned and closed to consolidate resources and reduce the overall sustainment/logistics footprint in support of the theater basing strategy. Activities include legal requirements; real estate reconciliation; real and personal property transfer and turn-in; reduction of contractual requirements; environmental cleanup; and records and archives to document transactions, agreements and conditions. Proper transfer and closure procedures facilitate the timely withdrawal of U.S. forces, reduce cost, prevent undue liabilities, protect U.S. interests, and promote good relations.

COMMAND AND CONTROL

1-44. Command and Control is the driving force throughout the base camp life cycle. It emphasizes the role of commanders at all levels in directing, leading, synchronizing, operating, managing, and continually assessing all aspects of the base camp life cycle to achieve effectiveness while improving efficiencies and conserving resources. A base camp has one directed commander for unity of command and unity of effort. The most senior commanders of tenant units, the senior airfield authority, other agencies, and contractors providing base camp support and Services must all work together—especially during security and defense operations, to ensure unity of effort.

1-45. Commanders foster base camp operational efficiency and effectiveness that yield adaptability and sustainability for meeting future requirements and rely on the commander's ability to deal with uncertainties in mission duration, troop levels, degree of permanence, political and civil conditions, funding to anticipate and manage transitions during the course of a campaign. Commanders at all levels work together in creating the context for base camps in the operational area, which includes—

- Ensuring the early anticipation and identification of base camp requirements for each phase of the operation.
- Conducting in-depth analysis of basing requirements and allocating the necessary resources and capabilities to subordinate units, through force tailoring and task organization, to enable effective planning, design, construction, and management of base camps at the lowest level. This is critical in mitigating the lack of organic base camp capabilities at the BCT/RCT level and below.
- Providing and enforcing the necessary guidance and policies on facility allowances, construction standards, and QOL that are appropriate for the situation.
- Establishing base camp management, which ensures effectiveness while optimizing efficiency and conserving resources.

Chapter 1

1-46. Commanders cannot exercise command of base camps alone. The support they need is enabled through the establishment of one or more of the following organizations that are focused on base camps and that become part of the base camp command and control structure:
- **Base camp management centers.** Base camp management centers coordinate, monitor, direct, and synchronize actions needed for establishing, operating, sustaining, and managing base camps within an echelon AO.
- **Base cluster operations centers.** Base cluster operations centers (BCOCs) are established to control several subordinate base camps that may be grouped together in a cluster for mutual support for sustainment or protection. BCOCs are most commonly found at brigade/regimental levels or higher headquarters and are similar to the base camp management centers in organization and function.
- **Base camp working groups.** Commanders at all levels may form base camp working groups by grouping select staff members who meet to focus on base camp planning or problem solving. Base camp working groups may conduct the initial base camp development planning until the necessary augmentation needed for adequate base camp development becomes available. When a base camp working group is established, the commander normally designates a group facilitator to focus the group efforts and prevent the duplication of effort. The group facilitator should brief the commander and staff on a recurring basis to maintain visibility and command emphasis on base camps. For the Army, the sustainment cell chief leads the working group.
- **BOCs.** The BOC is the base camp centralized management facility that enables the base camp commander/BOS-I to exercise authority and direction and facilitates the management of base camp functions, services, and support.

BASE CAMP PRINCIPLES

1-47. The base camp commander/BOS-I and staff use the base camp principles as a guide for analytical thinking. These principles are not a set of rigid rules, nor do they apply in every situation. They should be applied with creativity, insight, and boldness. These principles are—
- Scalability.
- Sustainability.
- Standardization.
- Survivability.

SCALABILITY

1-48. Scalability is the ability to tolerate population fluctuations and incorporate changes in the base camp mission, level of services, or force protection level, without the need for redesign. Solutions remain efficient and practical even when a base camp becomes larger or smaller. The size, composition, and positioning of forces are continuously adjusted based on mission requirements. Base camps must be able to accommodate these often unpredictable demands and remain responsive to the commander's needs.

1-49. Base camp facilities and infrastructure must be scalable to equally handle increases and decreases in population with the least amount of resources and effort. This is especially important during transitions in support of base camp closures and realignments and transfers of authority when base camp populations are essentially doubled.

1-50. Base camp plans, designs, materials, components, systems, construction methods, operational staffs, and communications systems should all be modular and scalable. Comprehensive scalable base camp solutions are integrated and developed at the joint and Service levels.

SUSTAINABILITY

1-51. JP 4-0 states that sustainability is the ability to maintain the necessary level and duration of sustainment/logistics support to achieve military objectives. This means that base camps must achieve and maintain effectiveness within the means of available resources (materials, labor, energy, and funds) without placing any unnecessary strain on existing sustainment systems. Sustainability is primarily achieved through the minimization of demand and the cost effective consumption of resources. Although these two methodologies are similar, the former is generally not appropriate for survivability, health, safety, and other aspects of Service member welfare.

1-52. Sustainability is broadly aimed at optimizing efficiency in base camps and in no way discounts the overriding requirement for operational effectiveness. The importance of ensuring the uninterrupted provision of essential base camp functions through redundancy in the systems and protection of critical infrastructure is acknowledged. While the possibility of fully incorporating this principle is related to the expected duration of a base camp (with a greater possibility in longer-duration base camps), it remains important to platoon and company, shorter-duration base camps since those camps could become long-duration base camps as the operation progresses.

STANDARDIZATION

1-53. The Standardization of base camp policies, Service standards, guidance, system solutions, standard designs, and construction provides consistent expectations for commanders and drives the repetitive use of proven best practices and TTP. It helps achieve a higher degree of sustainability, reliability, and efficiency. Standardization also reduces the uncertainty in meeting mandatory requirements and provides for more accurate estimates of materials, scheduling, and cost. Using standardized, scalable, and adaptable designs and construction, such as those in the Army Facilities Components System (AFCS), simplifies construction programming activities, improves early planning, and provides consistency in the application of levels of capabilities and in the resultant facilities and QOL on base camps. Standardization is achieved by enforcing base camp standards and guidance articulated in the CCDR's basing strategy, planning guidance, and design guides.

1-54. Standardization is also applied to procedures, organizations, training, and operations needed for managing base camps. Standardization helps to improve and sustain proficiency and readiness through the universal application of approved practices and procedures. It reduces the adverse effects of personnel turbulence associated with reassignments and facilitates interoperability between different organizations.

1-55. Standardizing designs and construction throughout the operational area eases repair and maintenance efforts by allowing for common stock of parts and supplies, which helps reduce inventories. It also reduces skill and training requirements for maintenance and repair workers. Custom-made designs can prove to be more costly and difficult to maintain and repair based on the future availability of parts, materials, and skilled labor needed. Therefore, facility and infrastructure designs are based on standard or traditional designs and constructed with standard or stock parts and materials that are readily available locally or through supply channels. This is an important consideration with design-and-build contracts since the original contractors who may have the know-how can change and O&M contracts eventually run out—leaving the burden solely on current and future base camp owners.

1-56. The Army Facilities Standardization Program is a formal process for developing Army standards and standard designs. Standard designs include drawings and specifications developed to ensure the application of sound engineering principles in the design process. UFC are a DOD-developed consensus on facility planning, design, construction, and O&M criteria for use by all Service components. The Army Facilities Standardization Committee has final approving authority for all UFC that affect Army standards. Army standards are listed in a table of mandatory criteria containing functional requirements necessary to complete present and future military missions. These Army standards are coordinated with Army functional proponents and approved by the Assistant Chief of Staff for Installation Management in coordination with the Army Facilities Standardization Committee.

SURVIVABILITY

1-57. A primary purpose of base camps is providing a protected location from which to project and sustain combat power. Base camps depend on the application of effective protection strategies, generally achieved by developing a comprehensive protection plan consistent with the principles of protection articulated in ADRP 3-37. Base camps must be equally prepared to protect against the effects of hostile actions, nonhostile activities such as fire, and environmental conditions such as floods and earthquakes.

BASE CAMP ACTIVITIES

1-58. Base camp activities provide useful constructs to aid in describing general areas of knowledge and in visualizing and characterizing the base camp operating environment. Base camp activities are interrelated and interdependent; each activity provides an action that mutually supports the others. The foundation of all activities is master planning. The base camp activities are—
- Master planning.
- Operations and maintenance.
- Protection.
- Sustainment/combat service support (CSS).

1-59. During mission planning, base camp activities help commanders and staffs organize the broad range of base camp requirements and the supporting information and tasks required for execution. Base camp activities are used in organizing people and equipment within base camp management centers, BOCs, BCOCs, and base camp working groups to facilitate the exercise of authority and direction and the management of base camps. By designating and using base camp activities, commanders have a means to support the base camp through collaboration, enabling planning and execution.

MASTER PLANNING

1-60. Master planning is a continuous process that evaluates factors affecting the present and future development and operation of a base camp. The evaluation forms the basis for determining development objectives and planning proposals to solve current problems and meet future needs. For base camps, master planning is accomplished primarily at the combatant command or service component command level. Primarily, engineers are involved with master planning efforts, with assistance from other branches or disciplines. This includes those activities necessary for enabling base camp functions, services, and support.

1-61. Master planning is an integrated strategy for the design, construction, and maintenance of required facilities and infrastructure that integrates base camp improvements for protection, QOL for residents, and efficiencies and effectiveness. Proper master planning enables scalable and sustainable base camps, conserves resources, and prevents wasted construction.

OPERATION AND MAINTENANCE

1-62. Many of the installation and O&M requirements within this area require technical expertise to ensure safe and effective operations. These requirements—especially on BCT/RCT or larger, more complex base camps—typically exceed the base camp commander's/BOS-I's organic capabilities and require augmentation or contracted support. On smaller camps, when augmentation is unavailable, base camp commanders/BOS-Is must rely on reachback to technical expertise residing in higher headquarters base camp management centers or support agencies and centers such as USACE.

Construction

1-63. Construction includes the tasks and activities needed for constructing the base camp facilities and infrastructure. Many of these tasks require technical expertise to ensure safe and effective operations. These requirements, especially on complex BCT/RCT base camps, typically exceed the base camp commander's/BOS-I's organic capabilities and necessitate augmentation or contracted support.

Facilities and Infrastructure

1-64. This activity includes all of the tasks and activities needed for maintaining, operating, and repairing the base camp facilities and infrastructure including—
- Structures.
- Utilities.
- Roads.
- Areas and grounds.
- Communications and network infrastructure.

PROTECTION

1-65. Protection is the preservation of the effectiveness and survivability of mission-related military and non-military personnel, equipment, facilities, communications security (COMSEC), and infrastructure. Base camp commanders/BOS-Is and staffs synchronize, integrate, and organize capabilities and resources to preserve combat power and mitigate the effects of threats and hazards. Protection is a continuing activity, integrating all protection capabilities to safeguard the force, personnel, systems, and physical assets. The keys to base camp protection planning are identifying the threats and hazards, assessing the threats and hazards to determine the risks, developing preventive measures, and integrating protection tasks into a comprehensive scheme of protection. Base camp protection includes the consideration of all of the protection tasks within the protection/force protection warfighting function articulated in ADRP 3-37. Fulfilling this functional requirement is a shared responsibility between operational and base camp commanders/BOS-Is.

SUSTAINMENT/COMBAT SERVICE SUPPORT

1-66. Sustainment/CSS activities provide service and support in two major areas: sustainment/logistics and base camp services. Base camp sustainment/CSS activities incorporate these areas and also provide general-purpose shelters and systems and support to Service members.

Base Camp Sustainment/Logistics

1-67. This base camp activity pertains to the sustainment/logistics support needed for sustaining base camp functions, services, and support. This sustainment/logistics support can be provided by any combination of the base camp commander's/BOS-I's organic or augmented capabilities, tenant units through support agreements, or contracted support. It includes—
- **Supply.** This includes all classes of supply needed to sustain base camp functions, services, and support.
- **Transportation.** This pertains to transportation needed to perform base camp functions, services, and support such as waste disposal, delivery of supplies, and shuttle services. On BCT/RCT base camps, the distance between unit areas and centralized facilities, such as dining facilities and post/base exchanges, may not be convenient for walking. In those situations, base camp commanders/BOS-Is may decide it is more efficient or cost effective over time—based on competing demands, fuel efficiency, and wear and tear on tactical vehicles—to acquire commercial or General Services Administration nontactical passenger vans or buses for use in a shuttle service.
- **Maintenance.** This pertains to the maintenance of commercial and General Services Administration nontactical vehicles and special equipment such as incinerators, generators, and passenger buses that are procured through local purchases and contracting to specifically perform base camp functions, services, and support. The maintenance requirements for these items often exceed the base camp commander's/BOS-I's organizational maintenance capabilities. Shortfalls in maintenance capabilities may be fulfilled through unit augmentation, support agreements with tenant units, and contracted support.

Base Camp Services

1-68. Base camp services make up a broad category of field services, personnel services, and other sustainment-related activities to a specified function of a base camp. These services are best provided from fixed facilities to improve operational efficiency and effectiveness and improve the overall QOL for base camp occupants. Base camp services include—

- Billeting (including latrine and shower facilities).
- Dining.
- Medical treatment, including medical, dental, and veterinary services.
- Laundry.
- Financial.
- Legal.
- Religious.
- Postal.
- Morale, welfare, and recreation (MWR) activities.
- Post or base exchange activities.

1-69. Authorization and types of base camp services are generally aligned with the designated level of services for the base camp and are detailed in the theater base camp standards and higher headquarters plans and orders. Base camp services may be provided by any combination of the base camp commander's/BOS-I's organic or augmented capabilities, a tenant unit, or contracted support. The base camp commander/BOS-I establishes policy on the provision of base camp services (such as authorized customers and hours of operation) that is supportive of higher-level policy and the commander's guidance and ensures the quality of services and level of support being provided through inspections and customer feedback. The base camp commander/BOS-I may designate facility managers to provide added focus on certain areas and facilitate base camp management.

ROLES AND RESPONSIBILITIES

1-70. The responsibilities for achieving efficient and effective base camps are not limited to any specific echelon. The required actions are handled at every echelon from policy decisions at the national or Service level, down to the base camp commander/BOS-I of the platoon base camps. Agencies and their associated roles and responsibilities include—

- **Army Assistant Chief of Staff, Logistics (G-4)/U.S. Marine Corps Deputy Commandant for Installations and Logistics.** Provides the consistent Service integration, management, and guidance on base camp solutions, standards for levels of services, and QOL. The Army G-4 integrates base camp operations and maintenance at the strategic level.
- **Marine Corps Systems Command.** Catalogues, standardizes, and provides the organic equipment to allow Marine Corps units to build base camps.
- **U.S. Army Installation Management Command (IMCOM).** Provides installation management expertise and best practices that are transferrable to base camp operations and maintenance. IMCOM supports its civilian employees who volunteer to augment deployable regional support groups in key installation management roles on base camps.
- **U.S. Army Materiel Command.** Catalogues, standardizes, and provides base camp systems.
- **U.S. Army Training and Doctrine Command.** Develops a standardized, comprehensive base camp Army training program. The MSCoE is the Training and Doctrine Command proponent for base camps and provides standards for survivability. The Sustainment Center of Excellence provides sustainment and sustainment/logistics support to develop largely self-sustaining base camps.
- **USACE.** Provides standards for construction, guidance on scalability, standardization and modularity, expertise on contingency standard designs, and management of the AFCS. Manages the worldwide power contingency contracts that provide power generation and electrical distribution services in conflict and disaster response locations. Also provides deployable augmentation teams to support base camps.

1-71. Establishing, operating, and managing efficient and effective base camps, regardless of size and purpose, is complex and resource-intensive—not only in terms of the labor, equipment, and materials needed, but also the command and staff efforts that are required throughout the base camp life cycle. The requirements for each aspect of the life cycle transcend staff functional areas and demand a combined arms, systematic approach that incorporates the expertise from maneuver commanders, engineers, logisticians, safety specialists, preventive medicine (PVNTMED) personnel, veterinarians, environmental officers, protection specialists, and other members of the staff.

COMBATANT COMMANDER

1-72. The CCDR integrates all aspects of the base camp life cycle at the operational level. The CCDR develops a contingency basing strategy for the joint operations area as part of his or her strategic estimates, strategies, and plans to accomplish the mission. The basing strategy translates national direction and multinational guidance into a concept that supports strategic objectives.

1-73. The CCDR may delegate the authority for base camp decision making to service component commanders or to commanders exercising Title 10 United States Code (USC) authority. Decisions are often made in consultation with the HN, subordinate commanders, and U.S. Department of State representatives.

1-74. The CCDR specifies the construction standards for the overall operation for facilities in the theater in operation plans (OPLANs) and operation orders (OPORDs), to minimize the construction effort expended on any given facility while assuring that the facilities are adequate for health, safety, and mission accomplishment.

1-75. Base camps are often collocated with military ports and airfields. The CCDR or the joint force commander (JFC) delineates responsibilities between the base camp commander/BOS-I, the military port commander, and the senior airfield authority to ensure unity of effort.

SERVICE COMPONENT COMMANDER

1-76. Service component commanders establish a staff engineer section with a facilities and construction department that manages engineering and construction within the AO under the appropriate USC responsibilities. This staff engineer section is responsible for developing the base camp and beddown plan for all Service personnel and equipment arriving in the area of responsibility (AOR). With guidance from the CCDR and the approval of the service component commander, the staff engineer section provides guidance on engineering and construction missions; establishes standards for construction; conducts coordination with the HN; participates in funding, utilization, and resourcing boards; and coordinates with the USACE or NAVFAC and the theater engineer command. The service component commander's responsibilities include integrating the legal, force health protection, and other aspects of environmental considerations provided from the respective areas of staff expertise. Service component commanders produce a Service level basing strategy that subordinate commanders use as the framework for developing their basing strategies.

ENGINEER STAFF

1-77. In some organizations, an engineer staff assists the commander by furnishing engineering advice and recommendations. Engineer officers and staff members may be assigned within an operations section (common among service component staffs), under a sustainment/logistics section (more common in joint staffs), under an engineer section (more common on combined staffs), or dispersed among numerous staff sections. In every case, a senior engineer staff officer is identified and must assume responsibility for coordinating the overall engineer staff effort.

Chapter 1

1-78. Engineer officers (or engineer staff) are primarily responsible for preparing the engineering portions of plans, estimates, and orders that pertain to base camps; participating on project approval and acquisition review boards and base camp working groups, as necessary; and coordinating and supervising specific engineering activities for which the engineering staff is responsible. The engineer assists the commander by performing a variety of functions to synchronize engineering capabilities in the operational area. They include—

- Planning and coordinating engineer support that uses military engineering units and contractors.
- Recommending policies and priorities for construction and real estate acquisition and for Class IV construction materials.
- Planning and coordinating the procurement and distribution of Class IV construction materials.
- Furnishing advice on the effect of base camp operations on the environment according to applicable U.S., international, and HN laws and agreements.
- Recommending construction standards.
- Standardizing infrastructure systems and design approaches.
- Identifying engineering support requirements that exceed funding authorizations and organized engineer capabilities.
- Furnishing advice on the feasibility, acceptability, and suitability of engineering plans.
- Coordinating with DOD construction agents and other engineering support agencies through appropriate channels.
- Coordinating the development of waste management plans.

LOGISTICS STAFF

1-79. The sustainment/logistics staff assists the commander by furnishing sustainment/CSS advice and recommendations to the commander and other staff officers; preparing the sustainment/logistics portions of plans, estimates, and orders that pertain to base camps; participating on project approval and acquisition review boards and base camp working groups, as necessary; and coordinating and supervising specific sustainment/CSS activities for which the sustainment/logistics staff is responsible. See ADRP 4-0 for more information. The sustainment/logistics staff assists the commander by performing a variety of functions to synchronize sustainment/CSS operations in the operational area. These functions include—

- Planning and coordinating sustainment/CSS that uses military sustainment/CSS units and contractors.
- Recommending policies and priorities for the procurement and distribution of supplies and materials.
- Identifying sustainment/logistics support requirements that exceed funding authorizations and organized sustainment/logistics capabilities.
- Furnishing advice on the sustainability, feasibility, acceptability, and suitability of sustainment/CSS plans.

BASE CAMP COMMANDER/BASE OPERATING SUPPORT–INTEGRATOR

1-80. The terms base camp commander and BOS-I are synonymous. When a BOS-I is assigned and command remains with the AO commander, the BOS-I is normally given specified governance authority over activities of the base camp. *Base camp commander* is an Army term and is the designation used for commanders of base camps occupied by U.S. Army units only. When multiple Services occupy a base camp, the CCDR will designate a lead Service and appoint that Service lead as the BOS-I. Each CCMD has criteria for determining lead Service designation on a base camp. The roles and responsibilities for the base camp commander and BOS-I are the same, regardless of Service.

1-81. The base camp commander/BOS-I is responsible for managing the day-to-day operations of the base camp. This includes the protection of the base camp and its occupants and the provision of base services and support.

1-82. For smaller base camps, the senior tactical commander is typically dual-hatted as the base camp commander/BOS-I. The AOR commander may delegate that role to a subordinate such as a deputy commander, executive officer, or subordinate unit commander residing on the base camp. For BCT/RCT base camps, the base camp commander/BOS-I is the senior tactical commander or the commander of a supporting unit—who may be junior in grade—tasked to manage the base camp. An Army regional support group is a specialized unit with the specific organization, equipment, and mission to manage support area base camps. Other units such as an Army maneuver enhancement brigade (MEB) may also manage support area base camps or base clusters.

1-83. In headquarters with staffs, from Army service component command to battalion level, the sustainment cell chief (G-4/S-4) is the staff integrator for base camps. The staff integrator organizes and facilitates a base camp working group with other select members of the staff. The base camp staff integrator is the staff lead for base camp-related issues. The primary focus is ensuring that base camp requirements and related information are coordinated throughout the staff and integrated into all aspects of mission planning. The staff integrator is the primary point of contact for supporting units or organizations tasked with supporting the development of base camps.

1-84. The G-4/S-4 will hold base camp working group meetings at appropriate times throughout the planning process to synchronize efforts and consolidate base camp-related information being generated and gathered from each staff member's respective functional area. As base camp-relevant information is identified, the staff integrator disseminates it to the appropriate staff sections and any units and organizations supporting base camp development planning for further analysis. They then determine operational effects from their perspective for inclusion in their running estimates/staff estimates to enable situational understanding (SU)/situational awareness (SA). Managing information, focusing on obtaining relevant information, and preventing information overload are fundamental to effective planning. The unit's planning SOP should describe the base camp working group's roles and responsibilities. The planning SOP should also describe who attends certain events during the planning process along with expected inputs and outputs.

COORDINATING STAFF

1-85. A coordinating staff supports the base camp commander/BOS-I in understanding, visualizing, and describing the operational environment; making and articulating decisions; and directing, leading, and assessing base camp operation. Coordinating staff officers are the commander's principal assistants who advise, plan, and coordinate actions within their areas of expertise or a warfighting function. Coordinating staff officers may also exercise planning and supervisory authority over designated special staff officers.

1-86. Coordinating staffs are typically organized into functional sections or areas. While staffs differ by echelon and unit type, all staffs include similar staff sections. The base camp commander's/BOS-I's grade determines whether the staff is a G staff or an S staff. Organizations commanded by a general officer have G staffs; other organizations have S staffs. Most battalions and brigades do not have plans or financial management staff sections. A chief of fires/force fires coordination cell (FFCC), a chief of protection/force protection officer, and a chief of sustainment are authorized at division and corps levels.

Personnel

1-87. The personnel section serves as the principal staff for all matters concerning military and civilian human resources support. Specific responsibilities of the assistant chief of staff, personnel (G-1)/battalion or brigade manpower and personnel staff officer (S-1) include manning, personnel services, personnel support, and human resources plans and operations. A key base camp role for this section is providing administrative support for non-U.S. forces, foreign nationals, and civilian internees. See FM 1-0 for more information on human resources support.

Intelligence

1-88. The intelligence section serves as the principal staff for providing intelligence to support current and future operations and plans. This section gathers and analyzes information on enemy, terrain, weather, and civil considerations for the base camp commander/BOS-I. The assistant chief of staff, intelligence (G-2)/battalion or brigade intelligence staff officer (S-2), together with the assistant chief of staff, operations G-3)/battalion or brigade operations staff officer (S-3), helps the base camp commander/BOS-I coordinate, integrate, and supervise the execution of information collection plans and operations. See FM 2-0 for more information on intelligence operations.

Operations

1-89. The operations section serves as the primary staff for integrating and synchronizing the base camp operation as a whole for the base camp commander/BOS-I. While the chief of staff (executive officer) directs the efforts of the entire staff, the operations officer ensures the warfighting function integration and synchronization across the planning horizons in current operations, future operations, and plans. Additionally, the operations staff (G-3/S-3) authenticates all plans and orders for the base camp commander/BOS-I to ensure that the warfighting functions are synchronized in time, space, and purpose according to the commander's intent and planning guidance. See ADRP 3-0 for more information on operations.

Logistics

1-90. The sustainment/logistics section serves as the principal staff for sustainment plans and operations, supply, maintenance, transportation, services, and operational contract support. The G-4/S-4 helps the base camp commander/BOS-I maintain sustainment/logistics visibility through sustainment operations and plans. See ADRP 4-0 for more information on the sustainment warfighting function.

Plans

1-91. The plans section serves as the principal staff for planning operations for the mid- to long-range planning horizons at echelon or division and higher. At echelons below division, this responsibility normally falls to the operations section. For a base camp, this section plays a crucial role in developing policies and other coordinating or directive products, such as memorandums of agreement. See ADRP 5-0 for more information on the operations process.

Signal

1-92. The signal section serves as the principal staff for all matters concerning network operations (NETOPS), communications, information services, cyber security, and spectrum management within the AO of the base camp. For a base camp, the assistant chief of staff for communications (G-6)/battalion or brigade communications staff officer (S-6) coordinates contractor and maintenance support for all NETOPS, information services, and electromagnetic spectrum management. See FM 6-02 for more information on signal support to operations. See appendix G for communications support network requirements.

Financial Management

1-93. The financial management support section serves as the principal staff responsible for all resource management and finance operations. The assistant chief of staff, financial management (G-8) is responsible for those operational financial management tasks supporting the theater. In coordination with the financial management center and through the theater sustainment command, the G-8 establishes and implements command finance operations policy. See FM 1-06 for more information on financial management operations.

Civil Affairs Operations

1-94. The civil affairs operations section serves as the principal staff responsible for all matters concerning civil affairs. The assistant chief of staff, civil affairs operations (G-9)/battalion or brigade civil affairs operations staff officer (S-9) establishes the civil-military operations center, evaluates civil considerations during mission analysis/problem framing, and prepares the groundwork for transitioning the AO from military to civilian control. The G-9/S-9 advises the base camp commander/BOS-I on the military operations effect on civilians in the AO relative to the complex relationship of civilians with the terrain and institutions over time. The G-9/S-9 is responsible for enhancing the relationship between Army forces, the civil authorities, and people in the AO. The G-9/S-9 is required at all echelons from battalion through corps but only authorized at division and corps. Once deployed, units below division level may be authorized an S-9. See FM 3-57 for more information on civil affairs operations.

Fire Support

1-95. The chief of fires (Army)/FFCC serves as the principal staff element responsible for the fires warfighting function at division through theater army. At brigade and below, the fire support officer/FFCC serves as a special staff officer for fires. This officer synchronizes and coordinates fire support for the G-3/S-3, who integrates fire support into base camp plans and operations. The chief of fires/FFCC also has coordinating responsibility over the air and missile defense and air liaison. See FM 3-09 for more information on fire support.

Protection

1-96. The chief of protection/force protection officer serves as the principal advisor to the commander on all matters relating to the protection/force protection warfighting function at division through theater army. The chief of protection/force protection officer has coordinating staff responsibilities for the chemical, biological, radiological, and nuclear (CBRN) officer; the explosive ordnance disposal officer; the operations security officer; the personnel recovery officer; the provost marshal; and the safety officer. Brigade and lower echelon headquarters are not assigned a chief of protection/force protection officer. Instead, the G-3/S-3 has coordinating responsibility for these staff officers.

Special Staff

1-97. The number of special staff officers and their responsibilities vary with authorizations, the desires of the commander, and the size of the command. If a special staff officer is not assigned, an officer with coordinating staff responsibility for the area of expertise assumes the functional responsibilities of a special staff officer. During operations, special staff officers work wherever designated by the commander. Some of the special staff officers critical to base camp development and/or operation are listed below.

Air and Missile Defense Officer

1-98. The air and missile defense officer is responsible for coordinating air and missile defense activities and plans with the area air and missile defense commander, joint force air component commander, and airspace control authority. The air and missile defense officer coordinates the planning and use of all joint air and missile defense systems, assets, and operations. An air and missile defense officer is authorized at the division, corps, and theater army levels. See FM 3-01 for additional information on air and missile defense operations.

Chemical, Biological, Radiological, and Nuclear Officer

1-99. The CBRN officer is responsible for CBRN operations, obscuration operations, and CBRN asset use. For a base camp, the CBRN officer is typically responsible for assessing weather and terrain data to determine environmental effects on potential CBRN hazards and threats, overseeing the construction of CBRN shelters, coordinating across the entire staff while assessing the effect of enemy CBRN-related attacks and hazards on current and future operations, and advising the base camp commander/BOS-I on CBRN threats and hazards and passive defense measures. See FM 3-11 for additional information on CBRN operations.

Engineer Officer

1-100. The engineer officer is responsible for planning and assessing base camp survivability operations. The engineer officer is involved in planning and operations with more than just the protection/force protection warfighting function. For example, mobility and countermobility are part of movement and maneuver, general engineering is part of sustainment, and geospatial engineering supports the intelligence warfighting function. For base camps, the engineer officer is responsible for synchronizing and integrating engineer operations; preparing the engineer portions of plans, estimates, and orders that pertain to base camps; providing real-time reachback to engineering knowledge centers and supporting national assets; and providing oversight of base camp infrastructure construction and maintenance. See FM 3-34 and MCWP 3-34 for additional information on engineer operations.

Explosive Ordnance Disposal Officer

1-101. The explosive ordnance disposal officer is responsible for coordinating the detection, identification, recovery, evaluation, rendering-safe, and final disposal of explosive ordnance. An explosive ordnance disposal officer is authorized at corps and division levels. See ATP 4-32 for additional information on explosive ordnance disposal operations.

Provost Marshal

1-102. The senior military police officer on the staff is typically designated as the provost marshal and is responsible for assisting the commander in exercising control over military police forces in the AO. The provost marshal is responsible for coordinating military police assets and operations for the command. The commander typically designates the provost marshal as a personal staff officer for law enforcement issues concerning U.S. military forces and U.S. personnel. This ensures appropriate sensitivity and security regarding criminal investigations and personal information. Each echelon down to brigade level has an organic provost marshal and staff element to integrate military police forces. The provost marshal's office is typically aligned within the S-3/G-3. In division and higher staffs, the provost marshal cell may be further assigned to the protection cell. Regardless, the provost marshal cell has significant coordination requirements with other staff elements to ensure that military police assets are employed properly and that military police capabilities support the commander's intent and stated requirements in an efficient and effective manner. A provost marshal is authorized at BCT, division, corps, and Army service component command headquarters. See FM 3-39 for additional information on the provost marshal.

Command Surgeon

1-103. A command surgeon is designated at every level of command. This Army Medical Department officer is a special staff officer charged with planning for and executing the Army Health System mission. At the lower levels of command, this officer may be dual-hatted as a medical unit commander and may have a small staff section to assist him or her in planning, coordinating, and synchronizing the Army Health System effort within his AOR. The command surgeon is responsible for ensuring that all Army Medical Department functions, to include all issues related to health facilities planning and management, are considered and addressed in running estimates, OPLANs, and OPORDs. Support for this task (including judgments on the sizing and capacity of medical treatment facilities) requires assigned staff with specific health facility planner (Area of Concentration 70K91) training. There may not be an Army health facility planning officer assigned on the surgeon's staff at every level of command. However, there is one Army health facility planning officer assigned to the—

- Medical command (deployment support).
- Medical brigade (support).
- Army service component command surgeon's office, and/or joint force surgeon's office, with additional reachback technical assistance provided by the U.S. Army Health Facilities Planning Agency.

1-104. Army health facility planners provide direct advice and input to the echelons above brigade engineering staff with regard to all health facility planning above the brigade and/or battalion aid station level to ensure appropriate alignment with the theater or AO medical concept of operations. See ATP 4-02.1 for additional information on health facility planning and management.

Judge Advocate General

1-105. Legal counsel ensures that contracts comply with U.S. law; DOD, service component, and CCDR regulations and policy; and as applicable, HN law and any agreements between the United States and the HN.

Base Camp Staff Integrator/Base Camp Working Group Facilitator

1-106. The sustainment cell chief or G-4/S-4 is the primary staff integrator for base camps. He or she organizes and leads a base camp working group with select staff to focus on base camp-related issues. The base camp staff integrator or the base camp working group facilitator serves as the lead on ensuring that base camp requirements and related information are coordinated throughout the staff and integrated into all aspects of mission planning. This individual also serves as the primary point of contact for supporting units or organizations that may be tasked with supporting the development of the base camp.

This page intentionally left blank.

Chapter 2

Base Camp Planning and Design

Planning is generally thought of as a fundamental organizing process to determine requirements and to devise and develop methods and schemes of action to solve a problem. Base camp planning in general is a detailed and methodical process by which the necessary actions are developed to support the commander's base camp requirements in response to a mission need in light of specified constraints with available resources for a specific purpose.

Where base camp planning is conducted at multiple levels in an operational theater from the strategic level (theater posture plan or a specific country basing strategy) down to the operational and tactical levels, for many necessary reasons, master planning is a detailed process of steps more specifically aligned to creating a unique and site-specific plan for individual base camps. Master planning is one of the most important responsibilities of a base camp commander/BOS-I. Master planning is a continuous analytical process that involves the evaluation of factors affecting the present and future development and operation of a base camp. This evaluation forms the basis for determining development objectives and planning proposals to solve current problems and meet future needs. Each task of the process builds upon the preceding task, providing a logical framework for the planning effort. This process provides a means for sustainable base camp development that supports mission requirements. This chapter explains the need for overall base camp planning activities and site-specific base camp master planning as it applies to base camps, which is accomplished through a comprehensive and collaborative planning process that results in a master plan for each theater base camp. Master planning provides the foundation for the base camp activities discussed in the remaining chapters.

Note. Do not confuse master planning with operation or mission planning. Operation planning is an activity that focuses on planning, preparing, executing, and assessing military operations. Master planning does not replace contingency and crisis action planning or troop-leading procedures at the tactical level for conducting unit operations. Master planning focuses on lasting development meeting present mission requirements without compromising the ability to meet future needs. The skills and experience required for master planning are most commonly attributed to general engineering. See ATP 3-34.40/MCTP 3-40D and ADRP 5-0 or MCWP 5-10 for additional details.

PLANNING PRINCIPLES

2-1. While the life cycle activities of planning and design are closely related, they are different. Site-specific planning and detailed master planning help establish requirements based on the intended purpose of a camp, the camp's location, specific terrain and environmental considerations, and threats. The allocation of resources for the purpose of base camp construction starts with the planning. Once a base camp is constructed, master planning continues throughout the life cycle of the base camp, forecasting and meeting the future needs of the base camp. The planning process is connected to design and is focused on enhancing established designs, adapting existing facilities for use, or creating new elements of the base camp to accommodate specific mission requirements that available organic assets or stock design plans are not capable of meeting.

2-2. The military contributes to national strategic planning for contingency basing through joint operational planning. Planners recommend and commanders define criteria for the positioning of base camps and thresholds on QOL within the levels of services based on expected mission duration and link those criteria to the achievement of the end state.

2-3. When infrastructure does not exist or does not meet the CCDR's basing strategy, new base camps need to be designed, constructed, and operated. To achieve a common standard of living and QOL at these base camps, master planners use standardized designs for individual facilities or entire camps. Base camp standardization generally improves the pace of construction while maximizing the availability of standard construction materials, which ultimately reduces the burden on sustainability.

2-4. Base camps are addressed as part of the JFC contingency planning that is performed in anticipation of specific situations that likely would involve the commitment of military forces. This facilitates the timely development and inclusion of more specific base camp policies and guidance (as part of crisis action planning for a certain situation) which ensure that base camps are efficiently used to meet mission requirements throughout the duration of an operation. See JP 5-0 for more information.

2-5. Another agency use of the military base camp, including any unique requirements, should be identified and considered before the agency joins an operation. Effectively integrating the interagency community contributes to the success of base camps, especially during theater shaping and during the stability and enable-civil-authority phases of an operation, when joint forces may operate in support of other U.S. government agencies. While supported CCDRs are the focal points for interagency coordination in support of operations in their AORs, interagency coordination with supporting commanders is just as important. At the operational level, subordinate commanders should consider and integrate interagency capabilities into their estimates, plans, and operations. Depending on the agency's capabilities and the threat, the agency may have a turn-key base camp provided to them by the military, be a tenant on a military base camp, or have a base camp turned over to them when it is no longer needed by the military.

2-6. JFCs should be prepared for base camps to support multinational forces. Multinational force commanders develop basing strategies and standards in multinational channels. JFCs coordinate these actions at the national level through established multinational bodies and at the theater strategic and operational levels.

2-7. CCDRs develop a basing strategy as part of their strategic estimates, strategies, and plans to accomplish their mission. They plan the arrangement, linkages, and sustainment of base camps. Basing strategies address the ends, ways, and means for base camps and are linked to the campaign plan, operational plans, and mission planning. Basing strategies—

- Articulate authoritative direction.
- Assign tasks, forces, and resources.
- Designate assumptions and objectives.
- Establish operational limitations (constraints).
- Establish thresholds within levels of services (basic, expanded, and enhanced).
- Define base camp policies, standards, and concepts to be integrated into subordinate or supporting plans to ensure that base camps are optimally efficient and effective.

2-8. The foundation for in-theater base camp planning is the basing strategy developed by the CCDR for the AOR. The basing strategy is mostly conceptual and includes some detailed guidance such as the overarching guidance on base camp standards, facility allowances, and the strategic vision for base camps (links and nodes) within the operational area. Commanders at each echelon provide added focus and detail that allow subordinates to develop base camps in support of mission requirements and to create desired effects.

BASING STRATEGY

2-9. A basing strategy translates strategic objectives into a physical presence. The overall arrangement of base camps throughout the operational area, their sustainment, and their linkages and interdependencies with other base camps, operational forces, and agencies are described in the CCDR's basing strategy for the joint operations area. The basing strategy may change over time, and some base camps will evolve differently than envisioned.

2-10. A basing strategy addresses how bases and/or base camps are used to enable access; extend operational reach; support line(s) of operations; support the generation of combat power; and support the operational, protection, and sustainment requirements of deployed forces. Included as part of the basing strategy are the base camps standards for such things as construction, QOL, design, environmental protection, and survivability that are tailored to a specific joint operations area (or region). In addition to operational and tactical considerations, some of the principal factors that are considered in formulating theater-specific base camp standards include—

- Joint and Service policies.
- International agreements and treaties.
- U.S. laws and regulations.
- HN laws and local customs and practices.
- The availability of indigenous construction materials.
- The availability and capability of the local labor force.
- Access to existing facilities and infrastructure.
- Access to land for an increase of base camp space.
- The availability of water from developed and undeveloped sources.
- Power and energy considerations.
- Climate and terrain effects on construction material characteristics and methods of construction.
- The availability of pre-positioned stocks and modular base camp sets.
- The ability to move construction resources into and throughout the operational area.

2-11. A basing strategy is developed by the CCDR as a product of operational art and design and is part of the theater strategy or security cooperation strategy. The basing strategy has mostly conceptual ends, ways, and means while containing some detailed guidance such as base camp standards. The basing strategy is reflected in guidance on base camps contained in plans and orders such as a theater campaign plan, a country plan, or a specific OPLAN or OPORD. The CCDR may develop an initial theater basing plan and revise it as the campaign progresses. See JP 5-0 for more information.

2-12. The terms dispersed and consolidated describe the two generally opposite approaches for arraying base camps. Each approach has strengths and weaknesses that require careful consideration:
- **Dispersed approach.** The dispersed approach, with very few large base camps and many smaller ones, requires more base clusters or hub-and-spoke relationships to improve the use of limited resources. It offers closer proximity to objective areas to allow for greater local engagement and requires a larger aggregate base camp footprint with more line of communications (LOC), and shorter hauls.
- **Consolidated approach.** The consolidated approach, with fewer large base camps, may be better suited for selected situations for units with good tactical mobility. Having fewer locations reduces the aggregate number of dedicated commanders and staffs needed for operating and managing base camps. It also allows security and defense efforts to be concentrated on fewer sites; however, it may allow enemies to focus their attacks and limit greater local involvement and contact. In general, this approach may be seen as more efficient; however, it may tend to be less effective depending on the operational environment.

PLANNING APPROACH

2-13. Effective base camp planning begins with the accurate identification of requirements for each aspect of the life cycle and the generation of supporting estimates and schedules for each phase of the operation. Estimates include the resources (people with the necessary skills, units or organizations with the necessary capabilities, materials, real estate, and money) that are needed to fulfill identified requirements. The planning process provides the framework for integrating the actions of the commander, staff, subordinate commanders, and others. Commanders and staffs use the planning process described in FM 6-0 and MCWP 5-10 to determine their requirements for base camps and integrate base camps within the concept of operations. See appendix B for information concerning the planning process. The base camp development planning process is described in EP 1105-3-1 as a process generally used for the actual development of base camps.

2-14. The planning process provides the framework for integrating the actions of the commander, staff, subordinate commanders, and others. Considerations for base camp planning and design activities must incorporate the base camp principles (scalability, sustainability, standardization, and survivability).

2-15. Predetermined staff members who have the functional responsibility for base camps or the base camp working group, will meet at appropriate times throughout the planning process to synchronize efforts and consolidate base camp-related information being generated and gathered from each staff member's respective activity. As base camp-relevant information is identified, it is disseminated to the appropriate staff sections and any units and organizations supporting base camp development planning for further analysis. They then determine operational effects from their perspective for inclusion in their running estimates/staff estimates to enable SU/SA. Managing information, focusing on obtaining relevant information, and preventing information overload are fundamental to effective planning. If a base camp working group is established, the roles and responsibilities of its members are described in the unit planning SOP. The planning SOP should also describe who attends certain events during the planning process, along with expected inputs and outputs.

2-16. Commanders and staffs incorporate collaborative planning to leverage the information resources and planning support services of higher headquarters, subordinates, and supporting units. Collaborative planning is the real-time interaction among commanders and staffs at two or more echelons that are developing plans for a particular operation. An example of collaborative planning might be BCT/RCT planners working together with Air Force planners for the design of an airfield on a base camp. Another example might be BCT/RCT staff members planning a base camp with a specialized base development team before deployment.

2-17. As the operation progresses, base camp planning continues to address all aspects of the life cycle in support of future plans and operations. Base camp requirements and the tasks necessary to fulfill them are synchronized primarily through integrating processes and continuing activities. Commanders and staffs monitor the efficiency and effectiveness of base camps and continuously make adjustments to reduce the sustainment/logistics footprint, conserve resources, and shape conditions for transitions. Adjustments include base camps realigning and closing, increasing or reducing levels of base camp personnel services, and improving efficiencies in base camp operations and maintenance. At the base camp level this is master planning, which is one of the base camp commander's most important responsibilities. A BOS-I may assume master planning responsibilities when multiple Service components share a common location, or as designated by a JFC. See ADRP 5-0 and MCWP 5-10 for more information.

MASTER PLANNING FOR BASE CAMPS

2-18. In simplest terms, a master plan is an evolving, long-term planning document. For example, there are master plans for policies, strategies, resource development, financial management, and research programs. For base camps, a master plan establishes a framework and key elements of a site reflected in clear vision and guidance. One of the primary purposes of master planning is to apply comprehensive planning strategies through facility and infrastructure development. This typically includes diverse activities such as planning, programming, designing, constructing, reusing, and performing real estate actions, entering public-private ventures, operating and maintaining, and disposing.

2-19. While master planning is normally directed toward permanent installations and real-property facilities, it is a useful tool for bases and base camps. DODI 4165.70 establishes the requirement for installation master plans. UFC 2-100-01 prescribes minimum requirements for the planning process and products. Although these requirements do not apply to overseas contingency operations and areas, the planning principles can and should be applied to base camp planning as part of an overall basing strategy. The single most important reason for this is that a base camp may evolve into a permanent installation at some point.

2-20. There is no single format for a master plan; however, a strong base camp master plan should address five key components:
- A site description of the BCT/RCT (or larger-scale) organization, including design principles that will shape the base camp growth.
- An identification of funding sources and strategies necessary to build the base camp.

- A development schedule identifying—
  - Which structures will be built first.
  - Which structures will be built later.
  - Which decisions will be made early.
  - Which decisions can evolve in response to the operational environment.
- Guidelines and standards for developing plans for the base camp structures and components.
- A vision statement describing the current and future purpose of the base camp.

2-21. Master planning is one of the base camp commander's/BOS-I's most important responsibilities. The base camp commander/BOS-I, supported by a team of staff members or a base camp master planning working group, develops a master plan that serves as a long-term blueprint for the implementation of improvements. The actual master plan for smaller base camps may be retained at its hub base camp BOC or BCOC.

2-22. Base camp master planning generally follows the process used for permanent installations outlined in AR 210-20 or UFC 2-100-01. However, base camp master planning has a shortened planning horizon and is often not planned with the same level of detail as a permanent installation.

2-23. Base camps are continuously improved to increase protection, enhance the overall QOL for residents, improve efficiencies and effectiveness, and ensure sustainability and adaptability for future requirements. Master planning provides an integrated strategy for the design, construction, and maintenance of required facilities and infrastructure at the best possible cost that enables scalable and sustainable base camps. Master plans also provide the commander's strategy for meeting the challenges associated with a base camp, to include but not be limited to antiterrorism (AT)/force protection; reduced manpower, resources, and useable land; base camp realignments and closures; and dependence on contracted support. Poor master planning can result in the inefficient use of resources and land, wasted construction, and inadequate base camps.

2-24. The CCDR establishes the policies and procedures for developing, approving, and implementing base camp master planning in the joint operations area. The requirements for master planning are linked to the theater basing strategy and detailed in subordinate unit plans and orders. Each base camp with a life span of 6 months or more has a master plan that is linked to a higher headquarters broader master plan. Theater guidance addresses archiving requirements related to supporting documentation.

2-25. Base camp master plans include land use development (base camp layout) maps or graphics, available land information (lease boundaries, explosive hazard areas, and environmental surveyed areas), and supporting construction project lists. Base camp commanders/BOS-Is use geospatial data and software applications to plot and show projected modifications to current facilities and infrastructure to enhance their master plans. Master plans should include goals and objectives that subordinates can translate into specific policies and actions. Goal statements should be specific enough to allow for meaningful interpretation by subordinates and for practical application. See appendix C for more information on land use planning.

2-26. A good master plan addresses uncertainties such as resource availability, base camp populations, mission duration, and evolving threats as well as the consequences of alternative options for efficiencies. Base camp commanders/BOS-Is identify indicators and triggers to shape master planning and facilitate decision making. Examples of indicators include increased frequency of mechanical breakdowns or failures in utility infrastructure (such as low water pressure or electrical breaker tripping), backlogs at waste collection points or disposal sites, and increased demands on water supply, which directly increases wastewater generation. Triggers include assigned thresholds on the demand for utilities and services based on capacity limits (for example, the camp laundry facility receiving 100 bags of laundry per day), outputs that indicate under- or overutilization (such as load percentages on generators and billeting occupancy rates), and projected timelines for raising levels of base camp services that are linked to operational phases or transitions. Base camp master plans must also address environmental effects and assess the risks of modifications to base camps. Some of the areas that require environmental and risk assessments include hazardous materials (HAZMAT) temporary storage areas, hazardous waste (HW) accumulation points, ammunition and explosives storage areas, fuel storage and refueling areas, firing ranges, and waste treatment and disposal facilities.

2-27. The base camp master plan is continuously reviewed, assessed, and adjusted based on new requirements such as changes in prescribed levels of services, expansions and reductions, evolving threats to the base camp, and changes in base camp functions and tenant unit operational requirements. A base camp master plan generally consists of the following:

- **Master planning digest.** This is the foundation planning component of the master plan. It provides the base camp commander's/BOS-I's vision, goals, and objectives for the management and development of the base camp. It describes the aim of the master planning effort, challenges and opportunities, and the road map (focus areas and priorities of effort) to achieving long-range goals for the base camp. It is not simply a summary of the base camp master plan. It also includes analyses and can serve as a decision support document. Base camp commanders/BOS-Is consider the base camp principles in formulating their digests.
- **Short-range component.** This component looks forward about 2 years and is linked to the established timeline for increasing the levels of base camp services and/or transferring or closing the base camp contained in the theater basing strategy. It focuses on camp improvements needed to provide the desired level of service and QOL. It includes time-phase project lists across the various funding sources that are available. It also addresses any environmental clean-up actions needed to support transfer or closure.
- **Long-range component.** This component looks forward more than 2 years and generally less than 10 years, which is the limit for semipermanent construction, and is linked to the established timeline for increasing the levels of base camp services and/or transferring or closing the base camp as contained in the theater basing strategy. It focuses on improving efficiencies, prolonging endurance, and ensuring sustainability. This component addresses base camp expansion or reduction that is linked to the theater base camp realignment and closure strategy.
- **Base camp design guide.** This is a design tool for standardizing sustainable energy and water efficiency, safety, environmental concerns, protection measures; promoting visual order; and building means and methods. It is more necessary for base camps that will likely be transferred or become permanent facilities/sites.
- **Capital investment strategy.** This is the base camp commander's/BOS-I's overall strategy for using and investing in real-property facilities, as resources and useable land become available, to support the base camp purpose and functional requirements. The level of detail of the master plan depends on the expected life of the base camp and the complexity of its facilities and infrastructure. Master plans for smaller, shorter-duration camps may only require simple sketches, such as presentation slides, as long as the necessary detailed information can be conveyed, while those for larger, longer-duration camps may include fully engineered construction plans based on completed surveys.
- **Master planning working group.** Effective master planning requires expertise in community development, environmental engineering, construction design, and many other areas. Units operating base camps generally lack the necessary, capabilities for performing effective master planning and must rely on individual augmentation or technical assistance from higher headquarters or reachback to supporting centers (such as USACE and NAVFAC master planning teams).

2-28. Base camp commanders/BOS-Is involve tenant units and organizations in the master planning process. Tenant units and organizations (military, governmental, and contractors) provide input into master planning to include the necessary designs and details needed to fulfill their operational requirements. Commanders/BOS-Is of base camps that are longer-duration or that have been designated to be transferred and become permanent facilities/sites in the HN should also include the considerations of the HN government and local populations. This ensures that designs and architecture are aesthetically pleasing and suited to the area.

Master Planning Process

2-29. The master planning process involves data collection and analysis that leads to the development of concept plans and finally to the definition of long-range plans for the physical development of the base camp. The process tasks are—
- Establish vision.
- Collect and analyze data.
- Develop goals and objectives.
- Develop and evaluate alternatives.
- Select the preferred plan.

Establish Vision

2-30. The guidance received from a senior command—joint, Service component, or higher echelon—is a key source of planning information. In some cases, it takes the form of specified or implied tasks or constraints. In other cases, it is general guidance that the senior commander wants to see applied when and where possible. The intent of the base camp commander/BOS-I provides the vision for how the base camp should look, how it should operate, and what the construction priorities are. Base camp planners integrate the base camp commanders/BOS-Is vision into the planning process and also advise the base camp commander/BOS-I on what is allowable, feasible, and efficient.

Collect and Analyze Data

2-31. Data collection and analysis are sequential steps. The task begins with an initial compilation of information covering a broad range of conditions. Analysis of this information identifies specific needs and significant constraints in meeting those needs. These opposing factors, needs, and constraints are then further analyzed to identify potential improvements. For a proper synthesis to occur, data collection and analysis must be thorough.

Develop Goals and Objectives

2-32. Specific goals and objectives for future development will provide guidance for developing the alternate concept plans and standards for the evaluation of alternate plans. The goals and objectives should address broad functional and location considerations for future physical development; they should not focus on specific facilities or sites. These goals and objectives also should be responsive to the operational and mission variables at the base camp.

Develop and Evaluate Alternatives

2-33. Alternate concept plans of the base camp depict the generalized long-range development, including the arrangement of land use areas, roads, and utility systems. The plan provides logical arrangements for the physical components of the base camp. They are derived from diagrams of ideal spatial relationships and adjusted to the reality of existing facility locations and the operational environment.

2-34. The number of alternate concept plans is determined by the possibilities for different logical arrangements of the base camp physical components. Generally, no more than three alternates are required unless a possible change in mission implies a wide variety of future requirements. However, in the initial stages of concept development, a range of alternates for a particular element (such as the circulation system or one or more utility systems) could be considered when there is no alternative available.

Select the Preferred Plan

2-35. The preferred plan is selected from the alternate plans. The advantages and disadvantages of each plan should be evaluated against one other and against the ideal spatial relationships. This evaluation determines the elements in each plan that come closest to achieving the ideal plan. One alternate plan may be selected as the most appropriate. However, the best elements of each alternate are typically combined to form the preferred plan.

Master Planning Products

2-36. The CCDR, typically through a joint engineer organization, provides guidance on master plans, including the required plan elements. At a minimum, the master plan should include the components shown in figure 2-1.

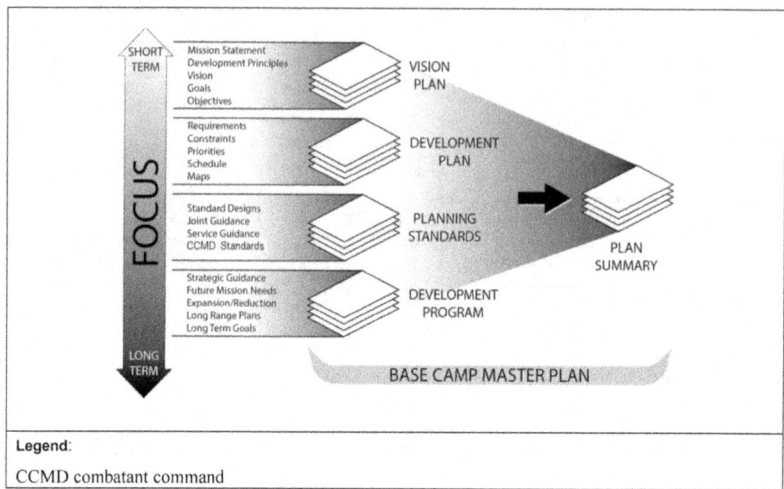

Figure 2-1. Base camp master plan products

Vision Plan

2-37. The base camp mission statement cites the specific responsibilities that the base camp must support. The vision plan is near-term and meets current military needs. Missions change as military requirements change. A vision for planning differs from an overall mission statement in that it defines ideal development principles for maximizing the long-term capabilities of the base camp.

Development Plan

2-38. This plan includes detailed constraints, opportunities, maps, illustrative plans, regulating plans, implementation plans, capacity analysis, and supporting sketches and renderings. It also includes network plans, which contain the plans for future development for the base camp as a whole.

Planning Standards

2-39. Planning standards provide a clear set of guidelines to ensure that the base camp vision and planning objectives for development are achieved, even if the mission changes. At a minimum, these include the applicable construction standards established in JP 3-34. Combatant commands also provide theater-specific standards. Standards contained in the UFC documents are intended for permanent construction unless explicitly stated otherwise.

Development Program

2-40. This program is the overall strategy for using and investing in real property to support missions and objectives. It describes the permanent comprehensive/holistic solution and short-term actions necessary to correct deficiencies and meet current and future mission needs using a method that ensures infrastructure reliability and contributes to sustainable development.

Plan Summary

2-41. A plan summary document includes the vision plan, executive summaries of the development plan, network plans, and a summary of the development program. This summary should be prepared after the above planning processes and products have been completed.

DEVELOPMENT PLANNING

2-42. The base camp planning process consists of several, not always linear, tasks. The final product is a complete, site-adapted base camp plan that provides a logical and documented solution for a base camp location, land usage, and facilities for the base camp that will support the users and the mission. The base camp development planning process is shown in figure 2-2 (derived from EP 1105-3-1.) The base camp planning tasks are—

- Preliminary planning.
- Land use planning.
- Location selection.
- Facility requirements development.
- General site planning.
- Design, programming, and construction.
- Maintenance and updating of plans.
- Clean-up, closure, and archiving.

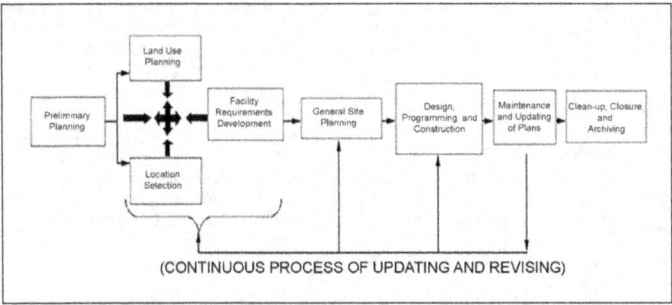

Figure 2-2. Base camp development planning process

2-43. These tasks are rarely performed in exact sequence; therefore, they are not numbered. Some of the preparatory tasks or actions may be performed concurrently or in advance to maximize the time available for planning. Some of the information needed, especially for preliminary planning, will be generated by the supported unit during mission analysis/problem framing. The sharing of this information in a timely manner helps to accelerate the process; however, that information must be reliable and based on facts and valid assumptions. Working ahead, even if based on valid facts and assumptions, is risky since the situation may change and assumptions may prove false, which can result in wasted time.

PRELIMINARY PLANNING

2-44. This planning phase involves gathering available information and determining information requirements. It resembles mission analysis/problem framing in the MDMP/MCPP in many ways. It may occur at any point in the operations process, although preferably during or after mission analysis/problem framing. If mission planning has already concluded, base camp planners should find much of the needed information in existing running estimates/staff estimates and staff products resulting from mission analysis/problem framing and IPB. Supporting units coordinate with the supported unit's base camp staff

Chapter 2

integrator, base camp working group facilitator, or a designated liaison officer to obtain base camp-relevant information resulting from the MDMP/MCPP. The preliminary planning tasks are—
- Determine possible base camp locations.
- Analyze base camp threats.
- Determine specified, implied, and essential tasks.
- Determine available assets and shortfalls.
- Determine facts and assumptions.
- Determine constraints.
- Determine time available.
- Determine information requirements.
- Begin risk management.

2-45. The use of the MDMP/MCPP processes for military planning makes the transition from mission to base camp planning an easy one. Appendix B provides a step-by-step application of the MDMP/MCPP process to base camp planning.

LOCATION SELECTION

2-46. The goal of base camp site selection is finding the best possible location for a base camp that balances mission, sustainment/CSS, protection/force protection, environmental considerations, and construction requirements. Site selection, the actual process in choosing a site, occurs later in the base camp planning process. Selecting the best location for a base camp is a balance between operational, sustainment, and construction requirements. It also involves consideration of the operational and mission variables. The selection of a base camp site occurs after the preliminary planning phase. See appendix B for site selection considerations.

LAND USE PLANNING

2-47. This task integrates the supported unit's requirements and the tenant unit's requirements, such as billeting, motor pool, storage, waste disposal, and protection needs, with land use affinities, operational constraints, and terrain restrictions. It provides a general overlay of land use areas within the proposed base camp. Since land use is directly affected by the selected site, this step is not finalized until the proposed site for the base camp has been approved. See appendix D for land use planning considerations.

FACILITY REQUIREMENTS DEVELOPMENT

2-48. In this task, planners must reconcile what is allowed with what is required. Facility requirements reflect the integration of facility allowances with unit requirements. Allowances are based on the type of unit, its size, and the anticipated life span of the base camp. These allowances are found in the theater-specific guidance documents and include areas such as allowable housing space, allowable command space, and allowances for specific facilities such as chapels and movie theaters. These allowances account for incremental or varying levels of service developed. JP 3-34 provides guidance related to facility standards. Once allowances have been determined, they are reconciled with specific unit requirements by validating or adjusting those requirements based upon specific unit needs. Adjustments to these allowances must be justified.

GENERAL SITE PLANNING

2-49. This task makes use of the initial land use plan, facility requirements, and coordination with unit requirements to complete the base camp design. The design includes individual building layouts shown within pre-identified land uses. Final decisions with regard to facility types, standards, construction, and the final location of specific structures and facilities are then made.

2-50. This task integrates the initial land use plan, facility requirements, and unit requirements into the base camp design. It includes individual building layouts shown within the predesignated land uses. Final decisions regarding facility types, standards, construction, and the final location of specific structures and facilities are made during this step. See appendix E for general site planning and layout considerations.

DESIGN, PROGRAMMING, AND CONSTRUCTION

2-51. The design, programming, and construction of base camps begin as early as possible. This ensures that funding and resources are available and that the base camp will be completed as scheduled. Beginning the design process early is essential in determining facility types and required resources and making recommendations for labor sources. It is also essential to ensuring that funding and resources are available and that the camp is completed in time to conduct its mission.

2-52. Programming for funds must be completed as early as possible to ensure adequate support. This is especially important if construction will involve the use of contractors, if lease payments are required, or if restoration and/or damage payments are anticipated. Certain funds may only be used for specific purposes. Base camp planners should consult with the supporting resource management office to determine fund availability, restrictions on use, and information on how to obtain funds and arrange for payments to vendors. If the project is congressionally funded, DD Form 1391, *FY__Military Construction Project Data*, is required. See AR 420-1 or ATP 3-34.40/MCTP 3-40D for more information.

2-53. The construction of key facilities such as those that support protection measures should begin as soon as plans are approved. Construction may be performed by military engineer units and contractors. Planners must determine, in coordination with those performing the construction, the proper sequence of events and the critical path required to execute construction in a timely and efficient manner. Planners must anticipate delays in shipments or deliveries of construction materials or services and mitigate the effects. Planners should be attuned to any critical items, especially those with long lead times. Planners must implement an effective quality assurance and surveillance plan, with qualified government construction inspectors, to ensure that contractors adhere to expected construction, safety, and environmental standards.

MAINTENANCE AND UPDATING OF PLANS

2-54. All construction projects require the maintenance and updating of construction plans. As these plans are altered according to the master plan, change drawings, diagrams, and environmental condition reports must be completed. An assigned integrator coordinates and synchronizes requirements and tasks between the master plans and construction plans based on the existing concept of the operations and future requirements. Contracts must specify the receipt of contractor-supplied plans for each portion of a project before payment for that portion or risk failure to capture the information. These plans are especially important where safety or environmental matters are involved. Additionally, any construction waivers that are requested and approved should be included with construction plans. Construction changes must be controlled according to the master plan.

2-55. An important part of transfers of authority is ensuring that the necessary base camp records and documents are handed over to the incoming base camp commander/BOS-I or retained in a central repository and readily accessible even as they are archived. Plans should initially be maintained by the base camp plans or operations staff or the base camp engineer. Theater guidance will provide further information on their final disposition. In all cases, the handoff of plans should be coordinated when units or responsible parties are changing (such as during unit rotations).

2-56. Master plans are communicated horizontally and vertically to all affected stakeholders. Subsequently, the stakeholders agree to the master plan or provide feedback. Lessons learned are continually captured as updates are made.

CLEAN-UP, CLOSURE, AND ARCHIVING

2-57. Planning for base camp transfers and closures early in the process helps avoid potential problems in the future. Depending on the situation, prerequisite actions needed for performing transfers and closures can be extensive. Prerequisite actions include environmental clean-up and restoration, the removal or destruction of facilities, the demilitarization of equipment, the turnover of facilities to the property owner or the HN, and the removal of materials. At the conclusion of a transfer or closure, key records that document the life span of the camp are archived to show what actions occurred; when, where, and how they occurred; and, in some cases, why they occurred.

FUNDING SOURCES AND AUTHORITY

2-58. Different types of O&M programs are identified in various sections of Title 10 USC that provide the authority for commanders to conduct operations. Normally, the supported and supporting CCDRs' service components will fund their participation in an operation with O&M funds. Contingency construction must satisfy the following four conditions:

- The construction is necessary to meet urgent military operational requirements of a temporary nature involving the use of the Armed Forces in support of a declaration of war, the declaration by the President of a national emergency under section 201 of the National Emergencies Act (50 USC 1621), or a contingency operation.
- The construction is not carried out at a military installation where the United States is reasonably expected to have a long-term presence.
- The United States has no intention of using the construction after the operational requirements have been satisfied.
- The level of construction is the minimum necessary to meet the temporary operational requirements.

2-59. Funding is a constraint that must be analyzed during planning. Military construction (MILCON) may be programmed or accomplished under a number of regulations and may be authorized and appropriated by separate acts of Congress. Typical funding sources for contingency construction are O&M, MILCON, and local purchasing. See JP 3-34 or FM 1-06 for more information on contingency authorities and funding.

BASE CAMP PLANNING AND DESIGN CONSIDERATIONS

2-60. Base camp planning identifies when, where, and why base camps are needed and the details of life cycle activities. Base camp planning begins as part of crisis action planning, is part of campaign and major operation planning, and continues through OPLAN and OPORD development and execution. See appendix B for planning considerations. Planning identifies the purpose and functional requirements of each base camp and linkages and interdependencies with other base camps, operational forces, and agencies, and generates the necessary information for executing all aspects of the base camp life cycle. It is linked to mission objectives and the commander's intent and results in a basing strategy and detailed guidance that directs the design, construction, and operations of individual base camps as part of a larger system of base camps.

2-61. Base camp planning activities cover a continuum that ranges from conceptual to fully detailed. Creating basing strategies or schemes of base camps involves mostly conceptual planning at the strategic and operational levels. Conceptual planning helps answer questions of what to do and why to do it. Conceptual plans are developed using base camp-specific assumptions to allow planning to continue despite uncertainty in a situation. Developing the basing strategy involves detailed planning at the operational and tactical levels. Detailed planning describes how to design a base camp.

2-62. The commander personally leads the conceptual component of planning. While commanders are involved in certain parts of detailed planning, they often leave the specifics to the staff and those individuals and organizations who specialize in base camp development. Base camp planning normally progresses from general to specific. The basing strategy, the conceptual component of base camp planning, provides the basis for all subsequent base camp planning and development. The basing strategy leads to schemes of base camps that, in turn, lead to detailed land use plans (site designs), facility and infrastructure designs, and construction directives. Base camp conceptual planning must respond to detailed constraints—for example, standards of construction, HN agreements, and available resources. These constraints are captured within base camp standards that drive the execution of the base camp life cycle.

2-63. Base camp development planning involves detailed planning. It translates base camp purpose and functional requirements into a complete and practical plan. Base camp development and mission planning have different focuses, while being interdependent. In combination, they ensure that base camps are positioned where they offer commanders the best means for projecting and sustaining combat power and where the terrain is favorable to engineering, design, construction, and environmental considerations. The fast pace of mission requirements during contingency operations rarely allows for these two planning methods to be conducted simultaneously, which complicates coordination and synchronization efforts. In some situations, base camps are planned as part of a sequel. This is often the case for a multiphase campaign involving major combat operations where the construction of base camps is largely contingent on the outcomes. Base camps require a flexible and adaptive approach to planning, just as all other portions of decisive action/simultaneous activities do.

PLANNING TEAM

2-64. Base camp planning requires a combined arms approach to harness the necessary expertise in the fields of sustainment/logistics, engineering, AT, protection, civil affairs, environmental resources, PVNTMED, resource management, safety, law, ranges and training areas, contracting, real estate, as well as other fields. It involves the unit staff of the primary organization that will be occupying the base camp, higher headquarters, and representatives from supporting units and organizations. Working together, they accurately identify the purpose of the base camp, its functional requirements, and the necessary supporting information early during the planning process. Based on those requirements, they work together in coordinating and integrating the necessary actions to fulfill those requirements. The management of data and information through common-access databases and shared networks is instrumental to this effort

2-65. The base camp commander/BOS-I may designate a staff member as the primary staff integrator for base camps and/or organize a base camp working group to focus on base camp-related issues. Although primarily focused on base camps, base camp working group members participate in every aspect of the planning process to ensure that base camp requirements and the supporting tasks are coordinated and synchronized within the concept of operations as it develops.

2-66. Depending on the scope of base camp requirements, a unit may receive augmentation from a specialized engineer unit, such as a forward engineer support team or an engineer facilities detachment, to assist with base camp development planning. When augmented, the supported commander may pass the lead on base camp planning to the supporting organization while using the base camp staff integrator or base camp working group facilitator to integrate the results of base camp planning into mission planning and ensure the sharing of relevant information. When the supporting unit or organization such as a base development team is not collocated with the supported unit, collaboration must be achieved through shared networks, voice and video conferencing, liaison officers, and other means of reachback. Commanders of the organizations ensure that the base camp planning being performed by their respective units or organizations remains mutually supportive through continuous coordination and information sharing. See appendix G for engineering reachback resources.

DESIGN CONSIDERATIONS IN PLANNING

2-67. General base camp design integrates functionality, protection aspects, base camp standards, and the prescribed level of base camp services to achieve sustainable and scalable facilities and infrastructure that fulfill the purpose and functional requirements of the base camp in the most efficient and cost-effective means possible. The purpose of the base camp is a primary driver in the design process. General design efforts in the planning process and detailed design for specific facilities are similar complementary activities.

2-68. There is no single correct design to a base camp. Each base camp has unique design characteristics based on site-specific criteria and the operational requirements of tenant and transient units. However, each base camp is also part of a broader system of base camps that is articulated in the CCDR's basing strategy and subordinate commanders' schemes of base camps. Base camp concept designs and approved detailed designs must comply with the basing strategy, standards, the master plan, and resource constraints.

2-69. Planners and designers consider the base camp principles (see table 2-1, page 2-14) within contingency construction designs to ensure efficient and effective base camps. The overall goals for base camp design are—

- Achieve functionality and sustainability while meeting operational and protection requirements.
- Adhere to established base camp standards and program and budget guidance.

Table 2-1. General design considerations in relation to base camp master planning principles

Base Camp Master Planning Principles	General Design Considerations
Scalability	<ul><li>Use modular and multifunctional designs.</li><li>Use modular buildings and trailer units that can be relocated, repositioned, and reused (or easily dismantled) and that offer flexibility.</li><li>Create designs that allow the base camp to easily expand or contract in size and levels of service.</li></ul>
Sustainability	<ul><li>Maximize the use of existing facilities and infrastructure.</li><li>Optimize the existing terrain characteristics.</li><li>Maximize the use of energy and water-efficient designs often termed *green designs*. This includes implementing shading and insulation whenever possible.</li><li>Use local resources (materials and labor).</li><li>Use energy and water-efficient equipment (generators, environmental control units, and low-flow toilets and showers) and materials (thermal insulation).</li><li>Maximize the use of hybrid and renewable energy sources (solar power) and reusable or recyclable materials.</li><li>Develop sustainable facilities and infrastructure (which are simple and inexpensive to operate, maintain, and repair).</li><li>Reuse or recycle energy and water (as applicable).</li><li>Minimize the use of spot generation since this typically results in generators that run under-loaded and inefficiently.</li><li>Maximize the use of smart-power distribution systems, and employ demand management.</li><li>Avoid siting facilities in low spots that are susceptible to flooding during the rainy season.</li></ul>
Standardization	<ul><li>Use standardized, scalable, and adaptable designs and construction.</li><li>Use standard systems and materials to simplify maintenance and repair.</li></ul>
Survivability	<ul><li>Optimize perimeter zone and entry control point alignment.</li><li>Provide appropriate spacing between structures.</li><li>Ensure an adequate standoff. (Position key facilities as far away from the perimeter as possible.)</li><li>Apply hardening where appropriate.</li><li>Construct a perimeter zone with supporting outer and inner security areas, including engagement area development considerations and other appropriate features and systems.</li></ul>

2-70. Base camp design is initiated as early as possible and in parallel with planning to ensure that planning and design remain mutually supportive and to provide adequate lead time on acquiring the necessary labor, equipment, and materials needed for construction. Critical information resulting from the design and integrated into planning includes construction estimates (bills of materials, equipment, personnel, cost, and time) that the commander needs to know in establishing priorities of support, priorities of effort, and timelines associated with the movement and basing of forces and the flow of the operation.

2-71. Planners and design engineers develop an integrated collection plan for base camp reconnaissance to support planning and design. Some information requirements needed to develop a basing strategy or early concept design may be obtained remotely. Most detailed designs will require on-site reconnaissance to determine conditions such as soil classification and the adequacy of existing facilities.

2-72. Any variables that affect design are resolved through planning. The primary variables include—
- The availability of suitable existing facilities and infrastructure.
- The availability of suitable construction materials and means for performing construction (skilled labor and special equipment provided by troops and/or contractors).
- Base camp standards (facility allowances and construction standards).
- The prescribed base camp level of services and linkages to other base camps as appropriate.
- Terrain and weather effects at the selected base camp location.
- Protection and security requirements (based on threat and vulnerability assessments).
- Civil and environmental considerations.
- Cost and time constraints.
- Governing U.S. regulations, policies, and HN laws and customs.

2-73. Base camp design is similar to planning; it is an iterative process that is continuously applied throughout the base camp life cycle and synchronized with other life cycle activities. See appendix D for additional details on land use planning and appendix F for facility and infrastructure design details.

ARMY FACILITIES COMPONENTS SYSTEM

2-74. The AFCS is the primary tool that provides engineers with the information needed to plan, design, and manage theater construction projects where austere, temporary facilities are required. AFCS and related series of publications provide a set of standard facility designs managed and supported by USACE. AFCS is an engineering construction support program for U.S. Army mission construction. Joint Construction Management System (JCMS) is the approved method for distributing AFCS designs and related information. The facilities and components in the AFCS typically satisfy many of the base camp construction requirements that are identified during planning. The AFCS facilitates base camp designs and can be used even in units at the lower tactical levels that typically lack the necessary design skills and capabilities. See TM 5-304 for information on the AFCS.

2-75. Suitable facilities and infrastructure designs (existing or established designs), such as those found in the AFCS, and prefabricated or pre-engineered buildings are used whenever possible. The AFCS provides designs for most of the facilities that are required on a base camp built to a temporary construction standard. These designs, in most cases, can be quickly site-adapted to suit the situation. The AFCS and Service doctrinal design and construction technical publications should provide metric designs and standards that can be used directly in those regions that use the metric system. Selecting designs early is critical to ensure adequate and timely resource availability. Creating designs that are already sustainable based on resources that are available is preferred. When new construction is required, planners use the detailed design process in developing and selecting facility and infrastructure design solutions.

2-76. The AFCS designs aid in solving problems associated with contingency construction constrained by resources and time. Facilities in the AFCS can be rapidly constructed with locally available materials. This allows for the use of preexisting supplies and indigenous craftsmen, which dramatically reduces costs and saves time.

2-77. The AFCS provides standard designs that are site-adaptable, scalable, and capable of serving multiple functions. AFCS temporary facilities are designed for the most typical and recurring types of facilities with an expected life of at least 5 years. Some of the facilities included in the AFCS are—
- Vehicle washrack.
- Warehouse.
- Medical treatment facility.
- Maintenance facility.
- MWR center.
- Joint operations center to include a sensitive compartmented information facility.
- Unit headquarters.
- Barracks.
- Dining facility.

This page intentionally left blank.

Chapter 3

Base Camp Construction

Whereas a base camp master plan often establishes the foundational details and overall scope of effort needed to begin constructing the base camp, it can also lead to unique site-adapted or mission-related requirements necessary for the development of specific, single-solution designs. Detailed design efforts may be necessary to determine how a particular facility, building, or supporting infrastructure element is surveyed and built. Design efforts are often required when it is necessary to modify existing designs, to renovate preexisting structures to create safe and usable facilities for our troops, or to create a new design to meet an identified need that current standard designs do not presently address. Construction itself is the art or process of building or assembling the structures that become the elements of a base camp, including beddown facilities or supporting infrastructure elements. Construction consists of a wide range of activities, methods, and techniques used to combine individual parts and to assemble resources to create a greater whole. Facilities and infrastructure are built using various methods that are determined and evaluated during planning and design. This chapter describes the detailed design process and general construction requirements, means, methods, and procedures for constructing base camps.

FACILITIES AND INFRASTRUCTURE

3-1. The base camp facilities and infrastructure necessary to support the force must be available at the base camp. This will often require new construction; where possible, it is important to maximize the use of existing facilities. Planning efforts should have included those facilities that cannot be sourced from existing assets. In these circumstances, when time allows, the appropriate Service, HN, or partner nation should perform construction during peacetime. Because construction is time-consuming and entails the risk of not being finished in time to meet mission requirements, engineers and planners should seek alternative solutions to new major construction efforts. Expedient construction methods with organic assets, prefabricated buildings or tents, are primarily considered, as these methods can be selectively employed with minimal effects on construction time, at a potentially reduced cost and with less risk to meeting mission requirements. Facility requirements are dependent on these factors:

- Mission and operational objectives.
- Total force structure to be supported.
- Expected duration of force deployment.
- Types of equipment to be employed.
- Number of days' worth of supplies to be stocked in the operational area.
- Standards of construction.
- Operational area medical policy.
- Operational area climatic conditions.
- Time-phasing of force deployment.
- Force protection.
- Availability and suitability of existing HN infrastructure.
- Real-property factors.
- Environmental restrictions.
- Cultural and historic sites and sensitive natural resources.

- Facility requirements factors.
- Safety requirements (explosive safety distances, airfield clearances, fire prevention).

Note. For more information see UFC 1-201-01.

CONTINGENCY CONSTRUCTION

3-2. The Secretary of Defense is authorized by 10 USC 2804, to execute contingency MILCON projects before a declaration of national emergency, upon a determination that the deferral of the projects would be inconsistent with national security or national interest. Specific funding limits can change each year. A staff judge advocate can provide guidance on current contingency construction limits and requirements.

3-3. In the event of a declaration of war or the declaration by the President of a national emergency requiring that use of the Armed Forces, the Secretary of Defense may undertake construction projects needed to support the Armed Forces without specific legislative authorization. Such projects, however, must be able to be completed within the total amount of unobligated MILCON funds, including funds appropriated for Family housing under 10 USC 2808.

STANDARDS

3-4. Construction standards are guidelines. The CCDR, in coordination with service components and the Services, specifies the construction standards for facilities in the theater to optimize the engineer effort expended on any given facility, while assuring that the facilities are adequate for health, safety, and mission accomplishment. Ultimately, the CCDR determines the exact construction type based on location, materials available, and conditions in theater.

3-5. Base camp construction standards are developed using operational and mission variables. Additionally, combatant commands consider the unique characteristics of the region and the anticipated duration of a mission in their basing standards. JP 3-34 provides joint contingency construction standards to be used as initial planning guidance. See UFC 2-000-05N for more information.

STANDARDIZED PLANS AND DESIGNS

3-6. To assist in the construction and establishment of base camps, DOD has established construction criteria and standard designs for many of the facilities and supporting infrastructure elements required for base camps. Many of these standards are found in the related UFCs and the standardized designs cataloged in automated tools such as JCMS.

3-7. The JCMS is an automated military construction planning system with a digital design, management, database, and reporting system used by military engineers for contingency construction activities in an operational area. It provides military planners, logisticians, and engineers with the information necessary to plan, design, and manage theater construction projects where austere, temporary facilities are required. JCMS is a tool that can be used for base camp development, planning, and design. Contingency construction plans and designs are generally characterized by—

- Speed of emplacement.
- Standardization, modularity, and scalability.
- Maximization of pre-positioned stock and locally procured standard building material.
- Ease of troop or available contract force constructability.

MISSION-SPECIFIC DESIGNS

3-8. The modification of preexisting detailed designs found in AFCS and JCMS may be necessary to alter the standard or existing design to fit existing site conditions found at the base camp location. Modifications can also occur to address new or varying design considerations that have occurred between the initial planning phases and the construction phase of the base camp. Encountering site conditions that differ from those initially planned or changes in the scope of the planned use of a facility are common and occur frequently, sometimes even after construction efforts on the base camp have commenced.

3-9. Reutilization of existing structures in a theater of operations requires various life safety code and structural inspections by qualified military or civilian personnel to determine the structure capability for continued/future use by U.S. or coalition forces. This ensures that all safety aspects have been fully considered before the occupation and use by U.S. Army, DOD Civilians, U.S. contractors, or coalition forces personnel. These structures often require redesign and construction modification to allow for a more optimized and safe use. For example, a billeting dormitory on an airfield may become the temporary coalition headquarters until such time as an improved facility can be constructed; to bring it to safety standards and to accommodate this intended use, for example, perimeter security, communications systems and backbones, and entry access control points must be constructed.

3-10. It may be necessary for the creation of new designs for facilities/structures when it is determined that existing facilities/structures, current standard/existing designs, or materiel kits (such as the elements found in the Force Provider System) are inadequate to meet essential mission requirements for the necessary base camp facilities and supporting infrastructure. New designs may also be required to address the availability of new construction materials or innovative construction methods made available for use.

GENERAL CONSTRUCTION REQUIREMENTS

3-11. A commander may be responsible for the planning, design, construction, management, and operation of a base camp; or a commander may only be responsible for constructing the base camp for another unit to manage and operate. In either case, planners, designers, and leaders within the constructing unit or organization consider operational and mission variables, available construction resources, theater construction standards, base camp levels of services, and the base camp principles in determining the optimal means and methods for constructing base camps. See the following publications for more information on construction planning and estimating:

- FM 3-34.
- MCWP 3-34.
- ATP 3-34.40/MCTP 3-40D.
- NTRP 4-04.2.3/TM 3-34.41/AFPAM 32-1000/MCRP 3-40D.12.
- NTRP 4-04.2.5/TM 3-34.42/AFPAM 32-1020/MCRP 3-40D.6.

3-12. Base camps are constructed in phases based on priorities and sound construction practices. Base camps are completed on time, within budget, and to the specified quality established in the approved design. When authorized and as required, the constructing organization adapts the design and finalizes the construction plan based on the actual resources that are available at the time of construction.

ENVIRONMENTAL, SAFETY, AND OCCUPATIONAL HEALTH CONSIDERATIONS

3-13. Commanders ensure that environment, safety and occupational health (ESOH) guidance and standards are being executed by the unit and/or contractors performing the construction. During construction, the generation of construction debris is a significant consideration and must be addressed in the overall waste management plan to minimize environmental effects. See ATP 3-34.5/MCRP 3-40B.2 and TM 3-34.56/MCIP 3-40G.2i for more information on waste management.

PROJECT APPROVAL PROCESS AND ACQUISITION REVIEW BOARD

3-14. Project approval processes and acquisition review boards ensure the equitable distribution of resources according to established priorities and prescribed base camp levels of service within the theater basing strategy. They also validate requirements against justifiable needs that are captured in base camp master plans, ensure best value, and prevent unnecessary or wasted construction.

3-15. A centralized project approval board and/or an acquisition review board is established at a predetermined echelon and meets regularly to provide oversight on base camp construction projects. The CCDR articulates the processes, procedures, and appropriate approval authorities based on identified thresholds for reviewing and approving base camp construction projects in the theater base camp standards. Base camp commanders/BOS-Is describe the procedures for project approval and acquisition review boards within base camp policies and SOPs and enforce standards with tenant units. A typical project approval process is shown in JP 3-34.

CONSTRUCTION PROCEDURES

3-16. The construction procedures used for base camps are normally executed by engineer construction units or contractors. The detailed construction procedures, techniques, and capabilities are found in Service doctrine for specific techniques or trades.

CONSTRUCTION MANAGEMENT

3-17. Base camp commanders/BOS-Is are responsible for establishing the appropriate means for managing initial construction or follow-on life cycle construction tasks on their base camps. This normally includes the appointment of project managers, who are responsible for the cost, quality, and timely completion of assigned projects.

3-18. Construction or project management involves six phases. These phases may be used for a single project or to manage multiple projects for the base camp. Figure 3-1 shows the typical construction project phasing model.

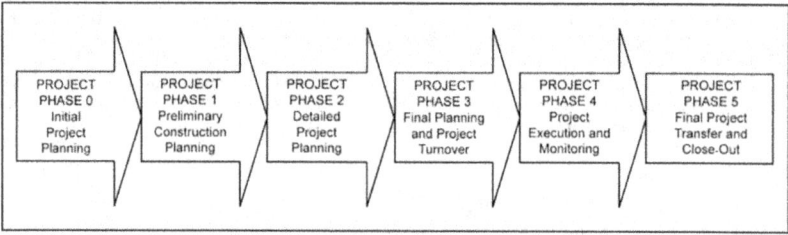

Figure 3-1. Construction project phasing model

3-19. Planners use the project management system described in NTRP 4-04.2.5/TM 3-34.42/AFPAM 32-1020/MCRP 3-40D.6 as a tool for coordinating the skill and labor of personnel using machines and materials for construction. The construction management phases are—

- **Initial project planning.** This phase begins when a sponsoring entity originates, sponsors, and funds a construction project. This phase may also begin with the receipt of a construction directive in the form of an OPORD, fragmentary order, or warning order for an engineering unit designated as the construction agent for a design-build project.
- **Preliminary construction planning.** This phase begins with the receipt of a construction directive in the form of an OPORD, fragmentary order, or warning order for projects that have already been designed. Preliminary project planning gives the engineering unit commander a quick overview of the assigned task, allows for an assessment of the unit capacity, and provides time to develop the commander's intent. It serves as a guide to the detailed planning that follows and includes the task assignment of primary personnel and acquisition of plans and specifications.
- **Detailed project planning.** This phase provides a schedule for the entire construction project and develops an accurate estimate of the tools, equipment, materials, and personnel required for each construction activity. It includes the creation or validation of a bill of materials; the completion of product and resource-leveling schedules; and the preparation and distribution of quality control, safety, and environmental protection plans through the chain of command for review.
- **Final planning and project turnover.** This phase includes project confirmation briefings, deployment/mobilization preparations, and the confirmation of initial material requirements availability.

- **Project execution and monitoring.** This phase includes the direction, monitoring, and implementation of corrective actions.
- **Final project transfer and close-out.** This phase includes turn-in of all tools and materials, clearance of the job site, provision of contractor-supplied records and operation and maintenance manuals, and the construction agent's receipt of final acceptance of the project. It may include submission of DD Form 1354, *Transfer and Acceptance of DOD Real Property*.

3-20. Project management normally begins with the development of a construction directive providing the who, what, when, where, and why of a particular project and generally follows the five-paragraph format used for mission orders and plans. See FM 6-0 and MCWP 5-10 for more information.

3-21. All construction projects require the maintenance and updating of construction plans. As these plans are altered, change drawings, diagrams, and environmental condition reports must be completed. When contract construction is conducted, the contract must specify the receipt of contractor-supplied plans for each portion of the project before payment is made for that portion or risk failure to capture the information. These plans are especially important where safety or environmental matters are involved. These include areas such as—

- Electrical systems, especially if buried lines are involved.
- Sanitation systems, such as buried sewer lines, sewage lagoons, and latrine pits.
- Ammunition and explosives storage areas.
- Training areas, especially those that produce explosive hazards.
- Landfills.
- Incineration sites.
- HAZMAT storage areas.
- HW accumulation points.

3-22. An integral part of the construction phase is reporting. Reports from the unit conducting the construction are used to measure and analyze performance, anticipate change and unforeseen requirements, and resolve problems. The headquarters directing the construction ensures that the necessary reporting requirements are contained in orders, plans, or SOPs.

CONSTRUCTION PLAN

3-23. The base camp construction plan describes who does what, where, and when. The when is based on established priorities, the arrival of construction units or contractors and the necessary equipment and materials, and the logical sequence for performing construction tasks. The base camp construction plan may also include other tasks not related to construction, such as real estate acquisition, funding, and security, which must to be performed before construction begins if those tasks are not covered in other plans and orders.

> *Note.* The base camp construction plan may be part of the base camp master plan. It may also be a separate plan developed for use by the constructing unit or contractor. Any plans developed for base camp construction should be maintained consistent with CCMD and Service guidance.

CONSTRUCTION PRIORITIES

3-24. Competing demands for often limited construction resources, including equipment, personnel, supplies, and funding, require careful prioritization. The basis for prioritization is largely a function of the importance of each project to the designated base camp purpose, function(s), and occupancy schedules. Priorities established in plans and orders may require adjustment after construction is started based on new mission requirements and changes in the availability of resources.

3-25. Base camp planners establish priority groups to facilitate decision making for resource allocation and master planning. See NTRP 4-04.2.5/TM 3-34.42/AFPAM 32-1020/MCRP 3-40D.6 for project prioritization.

Chapter 3

CONSTRUCTION SURVEYING

3-26. When a general area has been selected for a base camp, construction surveys are performed as time and the situation allow. Ideally, construction surveys are performed in conjunction with environmental and infrastructure reconnaissance and related surveys to obtain or verify the information required to finalize designs and begin construction. See ATP 3-34.81/MCWP 3-31 for more information. An on-site survey verifies information gathered from geospatial intelligence, including geospatial data and imagery that may have been used during planning and in making preliminary decisions. After completing a construction survey, the design information is transferred from paper to the field using construction stakes. Construction surveys include—

- Reconnaissance survey.
- Preliminary survey.
- Final location survey.
- Construction layout survey.

Note. See TM 3-34.48-1, TM 3-34.48-2, or AFPAM 10-219 Volume. 4, for more information.

3-27. The number of surveys conducted and the extent to which they are performed are largely dictated by the time that is available, the standard of construction desired, and the experience level of personnel performing construction tasks. The quality and efficiency of construction are directly proportional to the number and extent of surveys and other preconstruction activities. Key items to verify during on-site surveys include: threat situation, civil considerations, environmental considerations, existing site conditions, site layout, drainage considerations, soils classification, and the availability and quality of materials and the labor force.

3-28. Specific engineer technical expertise is required to conduct a survey. Survey teams can be augmented by forward engineer support teams, other technical specialties (such as medical, civil affairs, or other technical non-military entities), and reachback as needed to enhance the survey quality. See ATP 3-34.40/MCTP 3-40D for more information on construction reconnaissance and surveys.

Reconnaissance Survey

3-29. A reconnaissance survey provides a basis for site selection and provides information that supports future surveys. If a site cannot be selected based on this survey, it will be selected in the preliminary survey.

Preliminary Survey

3-30. A preliminary survey is a detailed study of a location tentatively selected on the basis of reconnaissance, survey information, and recommendations. If the best available site for a base camp has not already been determined, it is selected based on this survey. A permanent benchmark is established during the preliminary survey. Permanent benchmarks for vertical control and well-marked points for horizontal control are established. This enables construction elements to accurately locate and match specific design locations with those on-site.

Final Location Survey

3-31. A final location survey is conducted if time allows. Permanent benchmarks for vertical control and well-marked points for horizontal control are established. This enables construction elements to accurately locate and match specific design locations with those on-site.

Construction Layout Survey

3-32. A construction layout survey is the final preconstruction activity that occurs before construction begins. It provides alignments, grades, and locations (construction stakes) that guide construction. This survey continues until construction is complete.

CONSTRUCTION MATERIALS

3-33. Base camp construction uses materials that are versatile, durable, energy and water efficient, and readily available to enable timely constructed, cost-effective, scalable, and sustainable base camps. Versatile materials can be used and/or reused for various applications. Using materials that can be reused or recycled greatly reduces generated waste and waste disposal requirements. Materials must be durable and able to withstand the environmental conditions of the region.

3-34. Construction materials may be obtained using standard military supply procedures or local purchase procedures and contracting. Some materials may be available from pre-positioned stocks to facilitate rapid base camp construction. To maximize the benefits, local procurements should occur as close as possible to the base camp site, thus to minimizing transportation requirements.

3-35. Certain construction materials, such as fill soil, sand, gravel, and water, are often needed in large quantities. It is generally more cost effective to locally produce these materials through military or contractor-operated borrow pits, quarries, or wells. Consider how excavation will affect the surrounding community, land, and environment in terms of erosion, storm water runoff, natural habitat, and agriculture. Contracted construction and the construction directive for organic units should specify quality standards for the use of local materials that are verified through inspections as part of the quality assurance and surveillance plan. See ATP 3-34.40/MCTP 3-40D for more information on the production of construction materials.

SITE PREPARATION

3-36. Site preparation involves applying engineering designs and clearing the base camp site (or construction projects on the site) to facilitate the construction of facilities and infrastructure. ATP 3-34.40/MCTP 3-40D provides general details about site preparation. The following publications contain detailed information on site preparation requirements and techniques:

- ATP 3-34.80.
- TM 3-34.44/MCRP 3-40D.4.
- TM 3-34.45.
- TM 3-34.46/MCRP 3-40D.11.
- TM 3-34.47/MCRP 3-40D.3.
- TM 3-34.55.
- TM 3-34.56/MCIP 3-40G.2i.
- TM 3-34.62/MCRP 3-40D.9.
- TM 3-34.63.
- TM 3-34.70/MCRP 3-40D.5.

3-37. When earthwork estimation, equipment scheduling, and the necessary construction surveys are complete and worksite security is in place, construction begins by clearing, grubbing, and stripping. Clearing the land involves removing and disposing of all vegetation, rubbish, surface boulders embedded in the ground, and explosive hazards that may exist within the designated area. Grubbing consists of uprooting and removing roots and stumps. Stripping involves removing and disposing of objectionable topsoil and sod. These three operations are done primarily with heavy engineer equipment, power tools, explosives, and fire. The method of clearing depends on the—

- Amount of area to be cleared.
- Type and density of vegetation.
- Effects of terrain on equipment operation.
- Availability of equipment and personnel.
- Time available for completion.
- Threat situation.

3-38. For best results, a combination of methods is used in a sequence most suitable and effective to the operation. See TM 3-34.62/MCRP 3-40D.9 for more information on land clearing.

Chapter 3

Clearing and Grubbing

3-39. In most cases, engineer heavy equipment is the fastest and most efficient means of clearing and grubbing. Planners must evaluate the limitations on each type of equipment to be used based on the significance of obstacles in the area, such as the diameter of trees and stumps, and the effects of terrain on equipment operation based on surface configuration and soil conditions.

Stripping

3-40. Stripping consists of removing and disposing of the topsoil and sod that would be objectionable as subgrade, foundation under a fill, or borrow material. Examples include organic soils, humus, peat, and muck. Stripping is done concurrently with clearing and grubbing by using bulldozers, graders, scrapers, and front-end loaders. It is often helpful to stockpile suitable topsoil and sod for later use on bare areas for dust or erosion control.

Cut and Fill

3-41. Cut-and-fill tasks are conducted when clearing, grubbing, and stripping are finished. Cut-and-fill operations are the biggest part of the earthwork in base camp construction. The goal of cut-and-fill work is to bring the site elevation to design specifications. Throughout the fill operation, the soil is compacted in layers (lifts) to minimize settlement, increase shearing resistance, reduce seepage, and minimize volume change. Compaction is achieved with self-propelled or towed rollers. Cut-and-fill and compaction efforts are intended to achieve the final grade.

3-42. Depending on the size of the base camp, the amount of unsuitable materials can be voluminous and may require disposal. To reduce disposal requirements, efforts should be made whenever possible to reuse those cleared materials for other purposes. They should be used with caution as fill material in soil-filled containers. Coarse materials such as rock can become projectiles if the containers are subject to an improvised explosive device (IED) attack. Contaminated soil should not be used as fill material.

Drainage

3-43. A properly planned, designed, constructed, and maintained drainage system is essential to the serviceability of base camps. Delays caused by flooding, subgrade failure, and mud are avoided by employing an effective drainage system. Drainage structures should be developed in stages at the beginning of clearing, grubbing, and stripping operations to ensure uninterrupted construction. In most instances, surface water effects can be reduced by following the proper procedures for grading, compaction, and drainage. See TM 3-34.48-1, TM 3-34.48-2, or AFPAM 10-219, Volume 7, for information on drainage design.

3-44. Natural drainage features are used as much as possible to ensure the minimal disturbance of natural grades and to limit the necessary work involved. Where possible, grading operations should run downhill to improve efficiency and to capitalize on natural drainage. During clearing and grubbing operations, existing or natural watercourses must be kept cleared and holes and depressions filled. Adequate drainage for the site must be provided to ensure that water does not interfere with construction operations.

Soil Stabilization

3-45. Soil stabilization can be critical, even for short-duration base camps. Soil stabilization improves strength, controls dust, and renders surfaces waterproof. Even when the soil is not stabilized, it should be leveled and compacted to support the drainage of surface water, as required by the drainage plan; prevent soil saturation; and help minimize dust. See TM 3-34.64 for more information on soil stabilization.

3-46. Dust control alleviates or eliminates dust generated by vehicle and aircraft operations. Dust created by operations presents a health hazard and a hazard to equipment. Unfortunately, clearing large areas for motor pools, helicopter landing pads, roads, and billeting areas creates significant dust hazards. Various techniques to help suppress dust include placing larger-aggregate paved areas, ensuring that vegetative strips remain in place, and applying various chemical dust palliatives. Soil waterproofing maintains the natural or constructed strength of a soil by preventing water from entering it. See UFC 3-260-17 for supporting information. Additional techniques are presented in GTA 05-08-018.

3-47. Subgrades can be stabilized mechanically, by adding granular materials; chemically, by adding chemical admixtures (lime, Portland cement, fly ash); or with a stabilization expedient (sand-grid, matting, or geosynthetics). A stabilization expedient may provide significant time and cost savings as a substitute to other means of stabilization or low-strength fill. Matting and sand-grid expedients stabilize loose soil such as sand for unsurfaced road construction. Geotextiles and other geosynthetics are primarily used to reinforce weak subgrades, maintain the separation of soil layers, and control drainage. Geosynthetics are the primary means of waterproofing soils when grading, compaction, and drainage efforts are insufficient. The availability of these materials must be weighed with the considerable time savings for use of expedients in contingency construction.

CONSTRUCTION METHODS

3-48. There are multiple options for the construction of facilities and infrastructure that range from modifying preexisting structures; using pre-engineered metal or fabric buildings; using modular base camp kits; and constructing wood, steel, or CMU-framed and supported buildings. Construction technologies continue to evolve and offer improved methods of construction that may be incorporated through rapid fielding initiatives and contracted support to enhance the speed, quality, and sustainability of base camp construction. Standardizing the construction used throughout the operational area simplifies cost estimates, safety and quality assurance/quality control implementation, and facility maintenance and repair procedures; allows for reduced inventories in building materials and supplies; and reduces training requirements for construction workers. See ATP 3-34.40/MCTP 3-40D for more information.

EXISTING STRUCTURES

3-49. Existing facilities and infrastructure should be used whenever possible to save time, conserve resources, and reduce the overall sustainment/logistics footprint. Using existing structures assumes protection risks, in terms of survivability, safety, and force health protection, that must be mitigated through structural assessments and occupational and environmental health site assessments (OEHSAs). Documenting the existing environmental conditions helps limit liabilities. See UFC 1-201-02 for guidance on assessing existing facilities and ATP 3-34.5/MCRP 3-40B.2 for environmental considerations, assessments, and survey documentation requirements.

3-50. Using existing facilities is always a trade-off between protection and construction efforts. These facilities may not provide adequate protection from threats. Another consideration is that construction in foreign countries may not meet military or U.S. standards. For example, the construction could make extensive use of load-bearing walls. Existing facilities can be enhanced through retrofits, sidewall and overhead cover protection, and compartmentalization. However, if labor and cost exceed the protection benefits, it may be more advantageous to build a new structure. See ATP 3-34.40/MCTP 3-40D and UFC 4-020-01 for additional details)

TENTS

3-51. The use of organic unit tents or assembled packaged life support kits (if available) provide a quick means for establishing basic services. However, the effects of the long-term use of tents and the effects on protection and increasing services must be considered. The longer tents are used and exposed to the elements, the less likely they are to be easily repacked, stored, and reused. Tents used outside of the United States are typically not returned unless they can meet the rigorous cleanliness requirements directed by EO 13112. Additionally, tent costs, when combined with the costs of shipping into remote areas, may be higher than the cost of using local materials and labor to construct base camp facilities.

PRE-ENGINEERED METAL OR FABRIC BUILDINGS

3-52. Pre-engineered metal or fabric buildings, such as clamshell structures, are structures that are completely assembled on-site using standard components and materials brought to the site. They range from custom designs to packaged, preassembled and assembled kits, ready for construction. Some advantages they offer include rapid construction, flexible designs, durability, low maintenance, and minimal foundation preparation requirements. One of the major disadvantages is that the major structural components are often quite large and bulky and difficult to transport to the site.

MODULAR BUILDINGS OR TRAILER UNITS

3-53. Modular buildings or trailer units are types of facilities that are fabricated or assembled off-site, transported to the site, and placed in position. These structures come complete and are available in various sizes. They can be free-standing or placed inside an existing structure. Modular buildings may be used for multiple purposes and provide flexibility, speed of occupancy, and ease of expansion and relocation.

PREFABRICATED OR MANUFACTURED BUILDINGS

3-54. Prefabricated or manufactured buildings are structures that consist of several factory-built components that are assembled on-site to complete the unit. Prefabrication saves time on the construction site, which may be a factor when construction time is limited based on tactical or weather conditions. The construction site typically generates less waste, which reduces waste disposal requirements in the operational area.

TRADITIONAL CONSTRUCTION

3-55. Traditional construction using wood, steel, or CMU offers flexibility in designs, including the incorporation of necessary protection measures, and the ability to perfectly adapt to existing site conditions. Disadvantages include the time and effort needed for designing and constructing individual facilities, especially on a large scale. The environmental effects of procuring or using local construction material, (the depletion of the timber, soil degradation) must also be considered.

CONSTRUCTION MEANS

3-56. Sustainable base camps leverage construction resources that are readily available through local means, established supply channels, and operational contract support. See JP 4-10 and ATP 4-92 for more information on operational contract support.

3-57. Green or environmentally friendly construction materials should be used whenever possible. Green construction materials are characterized by such things as—

- Low toxicity (nontoxic or void of carcinogenic compounds and ingredients).
- Minimal emissions (low or no volatile organic compounds).
- Recycled content (produced with recycled materials).
- Recyclable materials (materials that are recyclable or reusable).

3-58. The cost effective use of materials and labor is achieved primarily by using local resources. Local resources are generally less expensive and avoid the challenges associated with international shipments; however, the quality of materials and services rendered must be considered in the overall cost-benefit analysis.

3-59. Given the fluidity of contingency operations, sustainment/logistics problems and labor shortages can occur with little warning. Where possible, anticipate and plan for delays and ensure adequate lead time to accommodate sustainment/logistics requirements.

3-60. Construction may be performed by joint and multinational engineer units or contractors, or a combination thereof that are balanced to meet established objectives that reflect mission requirements and the operational environment. Commanders must ensure that subordinate units and base camp commanders/BOS-Is are trained and capable of performing and overseeing the tasks needed for the construction methods established for the operational area. This training must also include the contracting officer representative (COR) training needed for ensuring the quality, completeness, and safety of contracted construction.

Troop Construction

3-61. The Army has baseline and specialized engineer units that can support base camp construction. Among these modular units are horizontal and vertical construction units, construction management teams, facilities engineer detachments, and survey and design teams. See ATP 3-34.22 and ATP 3-34.23 for more on engineer unit capabilities.

3-62. Vertical construction generally involves constructing, repairing, and maintaining protective structures (guard towers and bunkers), concrete structures, buildings, and associated utilities such as electrical, plumbing, water, petroleum distribution support, and sewage. The constructing unit may initially use organic capabilities to support its construction projects. If it is constructing a base camp ahead of planned occupancy, it may occupy completed facilities. It may need to construct support facilities separate from the base camp if another unit is scheduled to immediately use all completed facilities. These separate constructing unit facilities may be located within the base camp perimeter to efficiently use security and defense resources.

3-63. Horizontal construction generally involves constructing, repairing, and maintaining roads, airfields, heliports, drainage structures, pavement, and bridges. Horizontal projects may be completed by a horizontal unit or a unit with horizontal and vertical capabilities. Horizontal-engineering units augmented with support from vertical-engineering units prepare the site for the various construction projects.

3-64. The Marine Corps has engineering units organic to its Marine Air-Ground Task Force (MAGTF) elements that possess limited horizontal- and vertical-construction capabilities. These comprise the combat engineer battalions within the ground combat element, the engineer support battalions within the logistics combat element, and the Marine wing support squadrons within the aviation combat element. Each can plan for and construct base camps using organic expeditionary equipment possessed by the Marine Corps for mission-essential base camp facilities, and the engineer support battalion and Marine wing support squadron can construct expeditionary airfields. Additionally, as described in NTTP 3-10.1M/MCWP 4-11.5, the naval construction force (also referred to as Seabees) supports MAGTF contingency operations as an element and possesses a full array of horizontal- and vertical-construction capabilities to complement Marine Corps organic engineering capabilities and improve expedient, expeditionary facilities that are initially used for base camps. The Seabees augment the MAGTF with specialized capabilities not resident in the MAGTF, such as construction contract support, environmental specialties, design engineering, and public works management of base camps.

Defense Construction Agents

3-65. The *DOD construction agents* are USACE, NAVFAC, or other such approved DOD activity, that is assigned design or execution responsibilities associated with MILCON programs, facilities support, or civil engineering support to the combatant commanders in contingency operations (JP 3-34). The theater engineer command typically serves as the senior engineer headquarters for a theater Army, land component headquarters or, potentially, a joint task force. This command has responsibility for all assigned or attached engineer brigades and other engineer units and missions for the theater Army commander or joint force land component commander. When directed, the theater engineer command also provides command and control for engineers from other Services and multinational forces and oversight of contracted construction engineers. The command also coordinates closely with the lead DOD construction agent and senior contract construction agents in the AO. See JP 3-34 for more information on defense construction agents.

Contracted Construction

3-66. Use of construction contracts and contingency funding is important in developing base camps, especially those with expanded and enhanced QOL standards. At very austere bases, the lead Service responsible for base construction may use Service general engineering units (Army general engineering units, Naval Construction Regiment, and Air Force RED HORSE) to design, request and assist in the oversight of minor construction contracts normally executed through the Logistics Civil Augmentation Program (LOGCAP) task order or a theater support contract from the supporting Service contracting element. In the case of major construction, as defined in DOD policy and federal acquisition regulations, the lead Service will request support from the appropriate DOD-designated construction agent, normally USACE or NAVFAC. Unlike Service general engineering units, the USACE and NAVFAC have the full suite of organic design, contract award, and construction program management capabilities. Also, USACE can, when requested and within its limited capabilities, provide construction planning and advisory services for minor construction projects. See ATP 3-34.23, JP 3-34, and DODD 4270.5 for more information on contracted construction and construction agents.

3-67. Executing minor construction through contracted means in military operations requires significant management efforts from the contracting staff and the requiring activity and/or supported unit. The Service component commander ensures that the requiring activities are properly trained and actively participate in the planning, contract request, and contract execution process, including the training and certification of CORs. Minor construction requests using theater support contracts require detailed design-and-build parameters in the performance work statements (PWSs), accurate cost estimates, and detailed quality surveillance assurance plans—all of which require a professionally certified general engineer staff to prepare. Likewise, construction-related CORs must be qualified general engineers or arrangements must be made with the supporting contracting organization to provide qualified technical inspectors to perform the technical portion of contract execution. The property book office must establish and maintain accountability for all equipment procured or leased through contracting or local purchases.

3-68. A COR is a military or civilian government employee (not a contractor) appointed in writing by a contracting officer responsible for monitoring contract performance, through inspections and quality assurance checks, and performing other duties as directed (JP 4-10). CORs provide the technical knowledge, skills, and abilities needed to ensure that contractors are providing the desired products and services. Frequently, multiple CORs may be required for a single project. CORs play an important role during project initiation by helping to ensure the accuracy and completeness of the PWS. Incomplete or poorly written PWSs contribute to wasted efforts in terms of time and resources.

HORIZONTAL-CONSTRUCTION PROJECTS

3-69. Base camp horizontal construction projects may include ditches and earthen berms (for perimeter and ammunition/explosives storage areas), roads, parking lots, airfields, and heliports. The sequencing of horizontal and vertical projects is done as a part of project management. The individual base camp horizontal construction projects are normally phased and completed ahead of the vertical projects to provide access and site conditions to begin vertical construction work. Some horizontal projects, such as paving, may be completed after the vertical projects are completed.

COMBAT ROADS, TRAILS, AND ACCESS ROADS

3-70. A combat trail is a traveled, unsurfaced way that has been cleared of obstacles. A trail may be roughly graded by earthmoving equipment to provide a relatively smooth surface. Combat trails are usually adequate for tracked and wheeled combat vehicles.

3-71. Mobility is necessary for successful offensive actions. Base camps that support maneuver forces may require the improvement or construction of combat trails through areas where routes do not exist. Building combat trails and roads is a combat engineering task conducted in close support to ground maneuver forces that are in close combat and in support of mobility. Combat engineers are responsible for combat trails and roads.

3-72. The construction of the access roads to a base camp may be a totally separate project performed by a different construction unit or part of the base camp project. Haul roads may be temporary roads used to move construction materials by the shortest economical route to the base camp construction site. Access roads and haul roads are identified during planning and are scheduled for completion before or concurrently with base camp construction.

CONSTRUCTION SUPPORT FACILITIES

3-73. The constructing unit almost always requires horizontal- and vertical-construction support facilities that will require resources. Ideally, these facilities could be part of the final base camp and turned over to the base camp commander/BOS-I when the constructing unit no longer requires their use. The construction support facilities should be identified during planning and be scheduled to be completed before or concurrently with base camp construction.

CONSTRUCTION FORCE PROTECTION MEASURES

3-74. When occupying the base camp site, the constructing unit conducts reconnaissance, secures the site, and establishes initial job site security while also establishing the initial perimeter security. The perimeter security is phased, upgraded, and sequenced with the completion of required site work. The perimeter work may include the construction of ditches, berms, barriers, fences, protective positions, guard/observation towers, and fighting positions. The construction of protection measures is sequenced with the other horizontal- and vertical-construction projects. Some key infrastructure and facilities-hardening measures are integrated into the primary project as it is constructed. Potential projects include protection measures associated with—

- BCOCs and BOCs.
- Support facilities.
- Logistics sites.
- Troop concentration areas.
- Entry control points (ECPs).
- Vehicle checkpoints.

VERTICAL-CONSTRUCTION PROJECTS

3-75. Base camp vertical-construction projects typically include buildings, infrastructure, and utilities such as power generation and distribution systems; water purification and distribution systems; sanitation/waste collection/treatment/disposal systems; shower, latrine, and laundry systems; communication infrastructure; bulk fuel, water storage, and distribution systems; and structures associated with the base camp transportation infrastructure. The construction of buildings is normally phased to efficiently use the skilled construction work force. Buried-utilities projects are most efficiently integrated with horizontal site work. Infrastructure and utilities construction may be phased to provide temporary services to facilitate construction or allow the immediate occupancy and use of certain facilities. All construction must comply with applicable standards, such as uniform building codes, national electrical codes, national electrical safety codes, and military or HN standards.

This page intentionally left blank.

Chapter 4

Base Camp Operations and Maintenance

Base camp operations and maintenance link the systems of all base camp activities such as master planning and construction and transfer. Base camp operation refers to the operation and management of the physical base camp and the provision of base camp services and support measures. Base camp operations and maintenance provide the services, utilities, and protection needed to maintain the base camp mission and the base camp itself. This chapter explains base camp operations, management, organization, and capabilities that are needed to achieve the purpose of the base camp and to fulfill functional requirements.

BASE CAMP OPERATIONS

4-1. Base camp operation is the operation and maintenance of the base camp facilities and infrastructure and the provision of base camp services and support measures that are needed to achieve the purpose of the base camp and its mission. The BOC Operations Section is the nerve center of the base camp. Operations personnel coordinate activities and work directly with all other staff sections. The BOC operations section controls daily base camp operations, maintenance, and training.

Note. The BOC (or equivalent) ensures the warfighting functions—including sustainment/logistics—are synchronized in time, space, and purpose according to the base camp commander's/BOS-I's intent and planning guidance. In this regard, the BOC is analogous to the G-3/S-3.

4-2. Special considerations are required in situations where the base camp has an airfield. At the very least, a base camp has a helipad for medical evacuation and emergency resupply. A specialized aviation unit typically handles airfield operations. The BOC must integrate and synchronize daily air operations to include construction and repairs. See ATP 3-17.2/MCRP 3-20B.1/NTTP 3-02.18/AFTTP 3-2.68, TM 3-34.48-1, and MCRP 3-40D.2 for additional information on airfield establishment and operations.

4-3. Each BOC should be organized based on METT-TC/METT-T factors. While each BOC is organized differently, the typical responsibilities for base camp operation are—
- Administrative support.
- Utility services.
- Field services (QOL).
- Supply and distribution.
- Waste management.
- Facilities maintenance.
- Emergency management.
- Training support.
- Unit processing.
- Administrative support.

4-4. The BOC should serve as a centralized administrative office for the base camp. Base camp administration typically involves religious, medical, legal including Uniform Code of Military Justice, human resources, administrative, and financial services. Larger base camps may have individual sections or offices dedicated to these and other sustainment activities. See ADRP 4-0 for more information on sustainment activities.

Utility Services

4-5. The BOC provides or arranges for base camp utility services. These include, at a minimum, electricity, water, and sewage. U.S. military engineer and quartermaster units have the resources to provide initial utility services, which supplement organic capabilities. As a base camp matures, HN support is used when possible to free military units for critical missions. Support agreements with HNs require careful negotiation of the assistance provided.

Electrical Power

4-6. Electrical power is an essential element of military operations. Power and energy services are essential to minimize fuel consumption, which reduces sustainment/logistics support requirements and may provide cost savings within the overall base camp life cycle. A well-designed power distribution system provides continuous, reliable power. See ATP 3-34.40/MCTP 3-40D, TM 3-34.45, and TM 3-34.46/MCRP 3-40D.11 for additional information on prime power generation.

Energy Conservation

4-7. Cost-effective energy and power management is achieved through a holistic approach that includes implementing—

- Energy-efficient technologies such as generators, kitchen equipment, insulated buildings, and environmental control units.
- Equipment controls such as timers and occupancy sensors.
- Energy conservation programs that promote behavioral changes in consumers through awareness and leader action.
- Energy conservation plans.
- Micro power and smart power distribution grids.
- Renewable energy resources, as applicable, to reduce demands on fossil fuels and thus fuel resupply requirements.

4-8. Using smart power grids helps minimize the need for spot generation and allows for better generator utilization, which improves generator efficiency and reduces base camp O&M costs. Smart grids also allow for the effective integration of energy storage and renewable energy systems.

4-9. Leveraging energy conservation technologies and approaches may require technical and specialized skills or services for installation and O&M. The availability of these capabilities and the initial capital and O&M costs must be considered in a life cycle cost-benefit analysis of the expected duration of the mission and life of the base camp.

Water Purification and Supply

4-10. Water is a critical commodity. Establishing a self-reliant means for water production, packaging, storage, and distribution on-site or nearby allows base camps to shorten supply lines and reduces the overall demand on the theater supply and distribution system. It also reduces the number of required sustainment/logistics convoys and the inherent risks associated with them. Recycling gray water, capturing rainwater, and implementing and enforcing water conservation plans help reduce water demands. See ATP 3-34.40/MCTP 3-40D and ATP 4-44/MCRP 3-40D.14 for additional information on water supply, support and well drilling.

Sewage Treatment

4-11. Base camps with enduring capabilities likely use a waterborne sewerage system, as described in TM 3-34.70/MCRP 3-40D.5. If the system cannot be connected to an existing treatment facility, a treatment method must be developed. The two primary types of treatment methods commonly used during contingency operations are sewage lagoons and septic systems; however, a septic system with a drain field is preferred. See TM 3-34.56/MCIP 3-40G.2i for additional information.

Field Services (Quality of Life)

4-12. Field services help provide QOL for Service members in terms of feeding, billeting, medical care and hygiene services, and morale and welfare activities. The type and level of field services support provided at a base camp varies depending upon requirements and the existing infrastructure in the operational area. A variety of organizations provide field service support for base camps. See FM 4-40 for more information on field services.

Supply and Distribution

4-13. The BOC should serve as a centralized office for base camp supply and distribution activities, including warehousing supplies; determining and validating requirements; prioritizing requests; distributing, redistributing excess, and retrograding supplies. This includes not only supplies intended for the base camp itself, but also supplies for any tenant and/or transient units. Supplies are typically handled by quartermaster supply companies assigned throughout an AO. See ATP 4-42 and MCTP 3-40H for additional information.

Waste Management

4-14. The waste generated on a base camp places a significant demand on resources. HN municipal waste disposal or treatment facilities may be nonexistent, incapacitated, substandard, or beyond reach due to security or political considerations—placing the entire burden for waste management on the deployed force. Reducing this demand is primarily achieved by reducing generated waste. Employing the waste management principle of the three Rs (reduce, reuse, and recycle) is essential in reducing generated waste. Reusable water containers should be used rather than disposable plastic bottles, should be used to reduce generated waste and the added strain on the sustainment/logistics system, especially when local recycling is unavailable or cost-ineffective. See TM 3-34.56/MCIP 3-40G.2i and EP 1105-3-1 for additional information on waste management.

Maintenance and Repair

4-15. Maintenance refers to recurring, day-to-day, periodic, or scheduled work required to preserve a base camp facility for the effective utilization for a designated purpose. This includes work to prevent damage to a facility, which otherwise would be more costly to restore. Repair is the restoration of a base camp facility for the effective utilization for its designated purpose through reprocessing or replacing worn or damaged parts or materials, including parts and materials that are not corrected through maintenance.

4-16. Maintenance activities include reviewing maintenance service records, conducting physical inspections, ordering parts, requesting services, and scheduling the repair or maintenance of individual facilities or of the overall infrastructure. The BOC facilities and infrastructure section (or equivalent), under the guidance of the base camp engineer, typically manages maintenance and repair efforts. These activities may be done through a contract or by the base camps units.

4-17. All base camp maintenance personnel should be aware of the importance of repair methods and the timely recognition and correction of the basic cause of failure. Some maintenance is necessary based on normal wear and tear.

Facilities Maintenance

4-18. The BOC provides or arranges for general engineering capability to support facilities maintenance. The BOC must have a work order management system to control and manage maintenance and repair work.

Company base camps may use a manual log to track maintenance requests. Larger base camps may implement an automated request system. Regardless of the data management system, effective and efficient operations require all work, whether accomplished in-house or by contract, must be documented.

MAINTENANCE STANDARDS

4-19. Base camp maintenance standards are intended to keep facilities economically maintained and functional. Proper maintenance protects the U.S. government investment. Failure to maintain the facility may result in expensive repair costs. Each base camp must establish maintenance standards consistent with the type and expected duration of the facilities. Each base camp will be different; however some general standards include—

- Permanent, semipermanent, and temporary structures will be maintained to provide substantially the same capacity, efficiency, and standard of appearance and comfort for which they were originally designed, consistent with their planned use and life expectancy.
- Structural components, such as foundations, columns, trusses, structural frames, and connections, will be maintained regularly and in such a manner as to ensure preservation and stability of structures.
- Exterior and interior surface materials and finishes will be maintained to eliminate defects, to prevent damage, and to keep the facility in good operational and sanitary condition.
- The maintenance of all structures not scheduled for retention or transfer will be consistent with their planned use and economical life expectancy.
- Combatant commands and Service component commands use UFC 1-201-01 for maintenance standards of temporary and semipermanent facilities in-theater. Permanent facilities follow the UFC standards appropriate to the facility type.
- Smaller base camps may use a manual log to track maintenance requests. Larger base camps may implement an automated request system. Regardless of the data management system, effective and efficient operations require that all work be documented.

EMERGENCY MANAGEMENT

4-20. Commanders are responsible for establishing, directing, and controlling 24-hour emergency response to base camp incidents that are not necessarily the result of hostile actions. Commanders must ensure that their base camp is prepared for, can respond to, can recover from, and can mitigate an emergency. See TM 3-11.42/MCTP 10-10G/NTTP 3-11.36/AFTTP 3-2.83 for information on installation emergency management, which is applicable to base camps, and ATP 3-39.10 and MCTP 10-10F for additional law and order/military police considerations.

4-21. Base camp first responders should include medical personnel, firefighters, emergency facility and infrastructure repair crews, CBRN specialists, HAZMAT incident response teams, and military police. Base camps should be provided first responder, emergency lifesaving, and tactical and industrial equipment and required training (such as forcible entry tools, firefighting tools, and leak-sealing systems). Base camp assets may also be required to support foreign consequence management. See JP 3-41 for more information.

4-22. The base camp emergency management function may be conducted through the base camp commander's/BOS-I's unit command post or a BOC, which may be collocated or combined with a base defense operations center (BDOC) if one is established. The base camp commander/BOS-I may also designate an emergency manager to provide added focus on planning, preparing, and responding to emergencies. Risk management is the foundation of emergency response planning and should be completed prior to the development or update of the emergency plan. A base camp commander/BOS-I should develop contingency plans for hazards such as—

- Power outages affecting key facilities.
- Fires and explosions.
- Water main leaks and flooding.
- Fuel and other HAZMAT spills or leaks.
- Natural disasters (earthquakes, floods, hurricanes, tornadoes).

- Evacuations (complete or partial) and/or sheltering-in-place.
- Other hazards identified during the risk management process.
- Law and order considerations.

Fire Protection

4-23. Temporary structures generally use combustible materials. Austere environments often lack adequate water and maintenance resources to support modern fire suppression systems. Fires can result in the rapid loss of facilities and can spread quickly to other structures. An effective fire protection plan is critical to the safety of personnel, facilities, and equipment. Fire protection must be included in the design of base camps. This includes tent and building spacing, means of egress, wiring standards, use of flame-retardant materials, fire-fighting vehicle access, availability of a water supply, and fire protection and HAZMAT spill response equipment. See TM 3-34.30 and UFC 3-600-01 for more information.

Police Operations

4-24. Police operations encompass associated law enforcement activities to control and protect populations and resources to facilitate the existence of a lawful and orderly environment. Military police conduct police operations under three basic conditions: to maintain good order and discipline on bases or base camps, to establish and maintain civil security and civil control in support of the rule of law in an operational environment, and to assist local law enforcement agencies in times of crisis during defense support of civil authorities. See ATP 3-39.10 for additional information on police operations.

4-25. The criminal investigations division (CID) provides support for sustainment/logistics security, criminal investigations, and security of designated high risk personnel. Due to the limited availability of CID units, support is provided to the AO commander on an area basis, at each echelon. Base camp commanders/BOS-I request CID support through the Provost Marshal.

TRAINING SUPPORT

4-26. The base camp commander/BOS-I is responsible for ensuring that base camp management unit personnel are adequately trained and competent in all aspects of base camp operations. The BOC coordinates and schedules the use of training areas and ranges. See ADRP 7-0, MCTP 8-10A, and MCRP 8-10B for additional information on unit training.

UNIT PROCESSING

4-27. Tenant units occupy and reside on the base camp. Transient units and organizations come to the base camp for specified services and support, which may or may not include billeting. A BOC representative should meet with incoming unit representatives and conduct a briefing concerning base camp operations, policies, and unit responsibilities. Topics of the briefing may include—
- Check-in and check-out procedures.
- Medical treatment/emergencies.
- Billeting.
- Security.
- Site operations.
- Operating hours of support facilities.
- Vehicle parking.
- Base camp map.
- Environmental protection.

Chapter 4

BASE OPERATIONS CENTER

4-28. A BOC is the recommended centralized command facility for operating and managing the base camp. It is the base camp commander's/BOS-I's primary means for managing base camp functions and the provision of services and support to ensure efficiency and effectiveness. Key BOC tasks are—

- Conduct the operations process.
  - Plan.
  - Prepare.
  - Execute.
  - Assess. See ADRP 5-0 and MCWP 5-10.
- Monitor, assess, and manage base camp activities, services, and support.
- Plan and coordinate for contracted support.
- Coordinate with tenant and transient units/organizations, subordinate base camps (for base clusters), adjacent base camps, and higher headquarters.
- Plan and coordinate for emergency management (incident response and consequence management).
- Conduct master planning, to include land and facility space management.
- Direct and control base camp security and defense if a BDOC is not established.
- Perform base camp administrative tasks, to include record keeping.

4-29. The BOC is similar to a typical command post. Personnel and equipment are arranged to facilitate coordination, the exchange of information, and timely decision making. Well-designed BOCs integrate command and staff efforts by matching personnel, equipment, information systems, and procedures against their internal layout. See ADRP 6-0 or MCDP 6 for more information.

4-30. BOCs are organized into functional areas that generally reinforce the base camp to help focus efforts. A typical BOC organization is shown in figure 4-1.

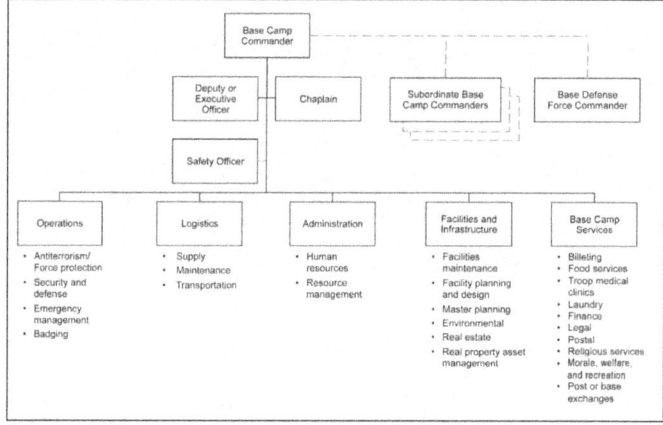

Figure 4-1. Example of a typical BOC organization

4-31. The BOC is staffed and outfitted using the base camp commander's/BOS-I's organic or augmented capabilities. Individual military or civilian augmentees and/or augmenting units are generally required for larger base camps where base camp O&M and management requirements are more complicated. For smaller base camps, with simpler and less-extensive facilities and infrastructure, base camp commanders/BOS-Is rely mostly on their organic capabilities. They assign additional duties [dual-hatting] and areas of responsibilities. They can also reach back to the expertise in higher headquarters (base camp management centers) and supporting organizations (USACE, NAVFAC) whose primary mission is to generate and sustain operational forces. The base camp commander/BOS-I of a smaller base camp may decide to establish a stand-alone BOC or assume the BOC function within the unit command post. The base camp commander/BOS-I develops a BOC SOP to explain individual roles and responsibilities, standardize procedures, ease the training of new personnel, and facilitate continuity during unit rotations. See FM 6-0 or MCWP 5-10 for more information about staff duties and responsibilities.

OPERATIONS

4-32. The operation section has functional responsibility for the access of training, operations and plans, and force development and modernization. The section also coordinates the intelligence, fires, and protection/force protection functions as directed by the base camp commander/BOS-I. The section may also provide input and/or coordinate base camp planning. For BOCs, the base camp commander's/BOS-I's S-3 or operations sergeant is typically dual-hatted as the base camp operations officer and leads the operations section. Key tasks performed within the operations section include—

- Planning and coordinating base camp protection, security and defense emergency management.
- Maintaining and updating the common operational picture (COP).
- Managing security clearances and security compliance.
- Establishing a badging and screening program for controlling access.
- Collecting, analyzing, and disseminating intelligence and combat information.
- Conducting IPB in support of base camp operations.
- Developing and managing the information collection plan.
- Conducting base camp assessments (risk, vulnerability, operational, resource, infrastructure).

LOGISTICS

4-33. The sustainment/logistics section function plans and coordinates supply, maintenance, and transportation tasks needed for sustaining base camp functions, services, and support. It is linked with the higher headquarters to ensure that base camp sustainment/logistics requirements (demands) are integrated within the overall plan for sustainment. The sustainment/logistics section plans and coordinates the following key tasks:

- Requisitioning, storing, and distributing base camp supplies.
- Maintaining base camp and commercial vehicles and special equipment used for base camp functions, services, and support.
- Transporting personnel, equipment, supplies, and waste (within the base camp and between base camps as part of a base cluster) as part of base camp functions, services, and support.

ADMINISTRATION

4-34. The administration section plans and coordinates base camp administrative tasks that may include military and civilian human resources support. This section may also have functional responsibility for resource management, identification badges, public affairs, religious services, medical services, and other areas as directed by the base camp commander/BOS-I. For BOCs, the base camp commander's/BOS-I's personnel staff officer or noncommissioned officer is typically dual-hatted as the base camp administration officer.

FACILITIES AND INFRASTRUCTURE

4-35. The facilities and infrastructure plans and coordinates the construction, maintenance, operations, and repairs of the base camp facilities and infrastructure. This also includes establishing and coordinating a base camp self-help program that defines the repairs, modifications, and construction that tenant units are permitted to perform. The section tracks the work performed to ensure compliance with safety and construction standards. All construction and modifications to facilities are coordinated with the base camp master planner.

4-36. Construction management is a key task within this function. The facilities and infrastructure function ensures that base camp utility maps are accurate and establishes and administers a permit process to track, verify, and authorize any excavation or earthwork to prevent damage to underground utilities. The function also coordinates and administers a facilities maintenance and repair work order system to ensure efficient use of resources, responsibility for environmental protection (see ATP 3-34.5/MCRP 3-40B.2), and master planning, as directed by the base camp commander/BOS-I. See TM 5-610 for more information on establishing preventive maintenance programs for buildings and structures, to include processing work orders.

BASE CAMP SERVICES

4-37. The base camp services function has responsibility for the broad categorization of field services, personnel services, and other sustainment-related functions that are provided as a specified function of a base camp. On smaller, short-duration base camps, this section may not be necessary and functional responsibility for the elements of base camp services is retained within the appropriate sustainment/logistics functional area. On larger, long-duration base camps, base camp services can be quite extensive and may require that an individual be designated as the base camp services coordinator or manager. The base camp services section is responsible for planning and coordinating the provision of base camp services. The base camp commander/BOS-I may designate facility managers for certain facilities. For example, a dining facility manager, may be designated to coordinate and manage dining facility activities.

BASE CAMP ENGINEER

4-38. The base camp engineer refers to the senior military or civilian engineer officer serving as the principal staff officer for the maintenance and repair of real property. The base camp engineer advises the base camp commander/BOS-I for all base camp engineering, operation, maintenance, and repair activities.

Note. Each Service refers to the installation engineer by a different title. The Navy uses public works officer; the Air Force uses base civil engineer, and Army uses base facilities engineer. Directorate of public works refers to a constituted U.S. Army garrison activity. Unless there is a formal directorate of public works organization on the base camp, refrain from using the term as it can be confusing and misleading.

BASE CAMP MANAGEMENT

4-39. A base camp resembles a permanent military installation in two respects. First, it has a mission aspect—a task or reason for existence. The mission may involve tactical, supply, administration, or communications (or any combination of these and other) activities. Second, it has a continuity aspect, sometimes referred to as base operation. This aspect addresses the physical property, the facilities, and the operation of the facilities to ensure the success of the mission.

4-40. The organizational structure needed for managing and operating base camps is fulfilled through base camp management centers and BOCs. BOCs are manned and equipped using organic and/or augmented capabilities. Augmentation for base camp master planning, design, program and project management, and real estate may come from USACE or similar Service capabilities. Augmentation for many areas of management and operation may come from IMCOM. Each BOC is tailored to meet mission requirements and organized to generally align with the base camp activities. The success of these organizations hinges on placing the right people with the right skills at the right place and time.

4-41. Individual augmentees and units, such as the Quartermaster Force Provider and specialized engineer or military police units, are generally needed at the higher echelons and for larger base camps where base camp O&M requirements are more complicated. Units establishing smaller base camps, at the lower tactical levels with simpler facilities and infrastructure, rely primarily on organic capabilities (dual-hatting), with the necessary expertise provided by higher headquarters (base camp management centers) and command-established assistance teams.

4-42. Expertise is also available through reachback to supporting organizations with a primary mission to generate and sustain operational forces, such as USACE, NAVFAC, and IMCOM. Any shortfalls in skills or capabilities at any echelon or any size of base camp, even those that may remain after augmentation, can be fulfilled through reachback. The skills needed for operating and managing base camps do not reside in any single branch or functional area. A grouping of capabilities is required to produce synergy.

BASE CAMP MANAGEMENT CENTER

4-43. Base camp management centers coordinate, monitor, direct, and synchronize actions needed for establishing, operating, sustaining, and managing base camps within an echelon's AO. Base camp management centers are typically established at division level headquarters and above, although based on the mission they may be formed within brigades and regiments if adequately resourced. They may be created using organic and/or augmenting individuals and units. Base camp management centers are similar to the functional and integrating cells formed within command posts. Personnel and equipment from select staff sections are organized to facilitate the accomplishment of mission objectives of the base camps.

4-44. The manning and configuration of base camp management centers varies between units and echelons, based on consideration of the mission variables. SOPs are developed to explain individual roles and responsibilities, standardize operations, and ease the training of new personnel. Typically, base camp management centers include operations, administration, environmental, base camp services, construction and design, and master planning staff sections.

BASE CAMP EMERGENCY MANAGEMENT

4-45. Commanders are responsible for establishing, directing, and controlling 24-hour emergency response to base camp incidents that are not necessarily the result of hostile actions—for example, power outages, water main leaks or flooding, fuel or HAZMAT spills, fires, and law and order issues. See ATP 3-39.10 for additional law and order information. Base camp first responders include medical personnel, firefighters, emergency facility and infrastructure repair crews, CBRN specialists; HAZMAT incident response teams, and (potentially) a provost marshal and other military police capabilities. Base camps provide first responder, emergency lifesaving, tactical, and industrial equipment, and the required training for personal which use them. Base camp assets may also be required to support foreign consequence management. See JP 3-41 for more information.

BASE CAMP CONTRACTED FUNCTIONS

4-46. Base support functions, including base camp services, construction, and facilities maintenance, often comes from commercial sources. This commercially provided support may include a mixture of the following types of contracted support LOGCAP task orders, USACE construction contracts, theater support contracts, and reachback contracts—the latter two executed by and through the supporting Army contracting support brigade. All base camp-related contract support, less commodity-only contracts, also involve a significant number of contractor employees, with a myriad of associated contractor personnel and management challenges.

4-47. Whatever contract mechanism is used, the base camp commander/BOS-I and staff often function with a base camp support-focused role with a critically important role to play in the planning, requesting and executing of operational contract support actions. More specifically, the base camp commander's/BOS-I's and staff will be responsible to plan for contract support based on theater-specific operational contract support, sustainment/logistics, and general engineering plan guidance. Depending on the type of contracted support that is used, this planning process may require the development of detailed contract support request packages to include PWS and draft quality assurance surveillance plans. In any case, the base camp commander/BOS-I and staff will be responsible for assisting in the contract administration process for this commercially provided services through the nomination of process-trained and technically qualified CORs. The inability to provide CORs with the proper technical qualifications to oversee these base camp services, construction and/or facilities contracts is a significant readiness issue that should be addressed through operational command and contracting channels as soon as identified. See ATP 4-10, ATP 4-92, and MCRP 3-40B.3 for more information on contract support.

4-48. Operational contract support can include significant contractor personnel management challenges. As stated above, commercially provided base camp services, construction, and facilities maintenance often require large numbers of contractor employees to work, and in some cases, live on the base. With this reality in mind, it is imperative that the base camp commander/BOS-I and staff include contractors authorized to live on the base, referred to as CAAF in doctrine and policy, in their personnel estimates. Additionally, base access and security procedures that will allow local national contractor employees to have access to their area of performance as directed in the terms and conditions of the contract must be developed and enforced. See JP 4-10 and ATP 4-10 and other applicable sections of this ATP for more information on contractor personnel management.

Chapter 5

Base Camp Security and Defense

Protection safeguards the base camp, its residents, systems, and physical assets from the effects of threats and hazards. Applying the protection tasks in ADRP 3-37 ensures that base camps provide the protection necessary for the critical camp infrastructure and personnel. This chapter focuses on base camp protection, primarily security and defense. Although generally aligned under the protection/force protection warfighting function, base camp protection integrates tasks from the protection/force protection and movement and maneuver/maneuver warfighting functions as articulated in ADRP 3-37 and ADRP 3-90, respectively. ATP 3-37.2, FM 3-90-1, GTA 90-01-011, GTA 90-01-034, JP 3-07.2, and JP 3-10 are additional key references for base camp protection.

BASE CAMP PROTECTION

5-1. All commanders are responsible for the protection of forces on base camps within their AOs. The base camp commander/BOS-I integrates the appropriate protection/force protection tasks as part of mission planning and throughout the operations process. The base camp commander/BOS-I implements protection/force protection measures for tenant units on the base camp.

5-2. Protection and defensive measures are applied within and beyond the confines of the base camp to safeguard personnel, physical assets, and information. Protecting and defending base camps include consideration of all of the protection principles within the protection warfighting function in ADRP 3-37/force protection in MCDP 1-0, and the associated defensive tasks detailed in FM 3-90-1 including—

- **Comprehensive.** Base camp protection is an all-inclusive utilization of complementary and reinforcing protection tasks and systems available to base camp commanders/BOS-Is, synchronized to preserve the force.
- **Integrated.** Base camp protection is integrated with other activities, systems, efforts, and capabilities associated with base camp missions to provide strength and structure to the overall effort. Integration must occur vertically and horizontally throughout the process.
- **Layered.** Base camp protection capabilities are arranged using a layered approach to provide strength and depth. Layering reduces the destructive effect of a threat or hazard.
- **Redundant.** Base camp protection efforts are often redundant anywhere that a vulnerability or a critical point of failure is identified. Redundancy ensures that specific activities, systems, efforts, and capabilities that are critical for the success of overall base camp protection have a secondary or auxiliary effort of equal or greater quality.
- **Enduring.** Ongoing base camp protection activities maintain the objectives of preserving combat power, populations, partners, essential equipment, resources, and critical infrastructure throughout the base camp life cycle.

5-3. Base camp security and defense capabilities are employed using a layered approach to provide strength and depth. Layering reduces the destructive effect from any single attack or hazard through the dissipation of energy or the defeat of the attacking force. A layered defense slows threat attacks and provides time for friendly defensive forces to assess, decide, and respond. Obstacles, such as barbed-wire fences, jersey barriers, T-walls, berms and ditches, bastion barriers, networked munitions, and direct-fire positions and elements are deployed in depth, in a concentric fashion, to provide maximum protection.

5-4. These obstacles, direct-fire positions, and active deterrents can be in the form of—
- Wire, concrete, or other barriers used to reinforce the perimeter.
- ECPs and associated obstacle/countermobility plans used to canalize and control incoming personnel or vehicles.
- Barriers employed to block high-speed avenues of approach, externally on approaches to the perimeter and internally to protect high-risk targets.
- Perimeter guard towers and observation posts (OPs).
- Ditches, berms, or other earthen obstacles.
- Mobile security patrols.

BASE CAMP THREATS

5-5. In most cases, base camps are located where the risk of Level III threats has been eliminated or effectively mitigated by the designated AO commander. However, base camps often become focal points for hostile actions. Because of the uncertainty in contingency operations and the acknowledgement of hybrid threats, all AO commanders must be prepared to conduct defensive tasks and designate a Tactical Combat Force (TCF) to repel a Level III attack when the threat assessment indicates the possibility of a Level III threat in the AO, regardless of whether the element of decisive action/simultaneous activities is currently dominant. Preparations may involve significant increases in area denial measures; offensive actions; hardening, dispersal, and other protection/force protection measures; and immediate reaction to hostile actions. While the hardening of facilities and the maintenance local security are the responsibilities of the base camp commander/BOS-I, area denial actions and offensive tasks designed to reduce the risks of Level III threats and to defeat those threats are the responsibilities of the AO commander.

Note. The term base camp commander/BOS-I refers specifically to a commander tasked with establishing and/or operating a base camp. There may be several base camps located within an AO. Some base camp command hierarchy exists, although relationships may be very different from standard Army doctrine. For clarity, AO commander refers generically to a commander with responsibilities for an entire AO. Under Army doctrine, only BCT and MEBs are assigned AOs. MEBs are typically assigned AO commander responsibilities for division, corps, and theater Army support areas; and could be assigned AO responsibilities for a joint security area. The AO commander, who may also be a base camp commander/BOS-I, is senior to all other local base camp commanders/BOS-Is within his or her assigned AO.

5-6. On initial occupation of the base camp site, friendly forces take offensive actions to identify levels of enemy presence and eliminate enemy threats in the immediate area, if required. Once the area is cleared and the necessary elements of the base camp defense have been established, the base camp commander/BOS-I continues managing area security tasks to provide an early warning and to mitigate the risks of threat elements operating within the base camp AO. The base camp commander/BOS-I and staff identify gaps in security and requirements for additional support or assets. The base camp commander/BOS-I, supported by the staff, coordinates with the AO commander to fill identified capability gaps.

5-7. Base camps are purposely designed and constructed to be resistant to enemy attack and to recover quickly so that they can continue to operate. The ability to quickly recover from an enemy attack is enhanced through detailed planning and rehearsals of procedures. Base camps must be prepared to defend in any direction through flexible base defense plans that include the use of dedicated initial-response forces positioned to respond to the widest possible range of contingencies.

5-8. The two principal types of attacks that a base camp commander/BOS-I and his or her staff focus on are categorized as penetrating attacks or standoff attacks. Infiltrated attacks from inside the base camp are likely to occur as well. Screening and vetting local workers are paramount to disrupting the threat's potential to gain access as a base camp worker or visitor.

Base Camp Security and Defense

PENETRATING ATTACKS

5-9. Defending against penetrating attacks relies on a strong perimeter defense that incorporates obstacles and integrated fires from well-protected firing positions. When applying defensive elements to a base camp perimeter, the type and extent of barrier and fires integration may be restricted based on mission and operational variables. Base camps within complex terrain, especially in support of stability tasks, will likely be restricted in the amount and types of obstacles and corresponding fires allowed in the outer security area; this is especially true for indirect fires.

5-10. Security forces must be capable of disrupting and delaying the penetration of the base camp perimeter until reinforced by an MSF or TCF. Base camp defenders should have tactical mobility with as much personal protection as possible. Security forces must be equipped with reliable and multiple means of communication. They should also have the necessary sensors and devices to execute reconnaissance and surveillance to the limits of the security area. This helps provide adequate detection and early warnings during periods of limited visibility.

5-11. Joint fires may be employed to augment the organic direct and indirect fire capabilities of the base camp. Security force personnel (augmentation and selectively armed personnel) may be directed to secure key facilities within the base camp, such as command posts, ammunition storage areas, and aircraft revetments. They may also support finding, fixing, containing, and defeating any attacks that may penetrate the perimeter. Adequate fire control measures must be employed to prevent fratricide.

STANDOFF ATTACKS

5-12. Standoff attackers are typically elusive targets. Level I and Level II threats may rely on blending in with the legitimate populace, only revealing themselves as combatants when they engage in a hostile act. Standoff attacks are mitigated by conducting area security tasks within and beyond the base camp AO to—
- Deny hiding places to the enemy.
- Disrupt enemy planning, reconnaissance, and organization.
- Detect the enemy as it moves into position and posture forces to quickly neutralize detected forces.

5-13. These preemptive actions rely on actionable intelligence, including human intelligence, within the base camp outer security area and beyond. For imminent threats that originate outside the base AO, for which the AO commander is unable to assist, the base camp commander/BOS-I must use organic base camp combat power to counter the threat or, with the permission of the AO commander, assume the risk of enemy standoff attacks.

BASE CAMP PROTECTION FRAMEWORK

5-14. The framework for base camp protection consists of three primary areas: an outer security area, a perimeter zone, and an inner security area. These are shown conceptually in figure 5-1, page 5-4. The base camp commander/BOS-I applies this framework to focus his or her planning activities and ensure that all critical elements of base camp security and defense are addressed. The framework is not intended as an all-inclusive solution to base camp protection; it is intended to provide a general template for planning.

5-15. Collectively, these three areas form the base camp AO. Commanders assigned an AO have inherent responsibilities that are described in FM 3-90-1. In those situations, the AO commander must clearly articulate that the responsibilities of the AO will not be performed by the base camp commander/BOS-I (and name who will perform the tasks), or provide the necessary augmented capabilities to perform the tasks.

OUTER SECURITY AREA

5-16. The outer security area is the area outside the perimeter that extends to the limit of the base camp commander's/BOS-I's AO. Commanders establish an outer security area to provide early warning and reaction time and deny enemy reconnaissance efforts and vantage points for conducting standoff attacks.

Chapter 5

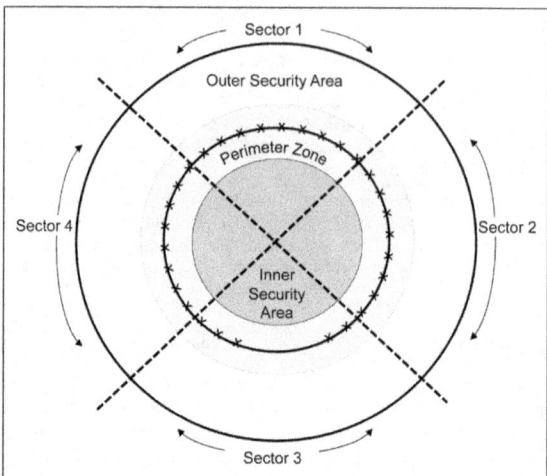

Figure 5-1. Framework for base camp security and defense

5-17. Each base camp and selected base clusters should have an outer security area that extends beyond the perimeter of the base camp. The outer security area limit of the base camp is the boundary that defines the extent of the base camp commander's/BOS-I's AO. The AO is established by the AO commander and is coordinated with the base camp commander/BOS-I based on consideration of the mission variables. It is a balance between the AO commander's and the base camp commander's/BOS-I's area security requirements and the necessary control of specific terrain to accomplish their respective missions.

5-18. The base camp boundary typically extends beyond the perimeter to at least direct-fire range (potentially out to the limits of the indirect-fire systems of the base camp), allowing base camps to execute their fire plans within their ability to control them. When possible, the outer security area should extend beyond the maximum range of any threat weapon that may be used against the base camp, including small arms and crew-served weapons, mortars and other indirect-fire weapons, and explosive devices that may be emplaced or activated by threat elements. Base camp outer security areas tend to be constricted by physical structures when base camps are established in complex terrain. See FM 3-90-1 and ATP 3-90.8/MCTP 3-34B for more information on the establishment of a security area.

5-19. The outer security area is typically patrolled by an MSF, typically military police. MSFs are a critical element in establishing a viable outer security area. Commanders ensure that detailed planning and coordination are conducted between security forces operating internally and on the perimeter and MSFs operating outside the perimeter to mitigate the risk of fratricide. Base camp commanders/BOS-Is must be prepared to conduct battle handover to a TCF.

5-20. The HN may limit the ability to conduct security tasks beyond the limits of the base camp perimeter, and may be capable and willing to assist with such tasks. In these cases, close coordination with HN security forces (HN military police, or other security elements) is conducted regularly to avoid fratricide and potential friction between coalition forces and the HN. The inability to adequately defend the base camp because of the HN limitations on the base camp AO must be communicated to the AO commander to revise existing agreements or provide acceptable risk mitigation.

5-21. The base camp commander/BOS-I conducts security tasks to the limit of the base camp AO. This is done to—
- Reduce uncertainties about the terrain and the enemy.
- Gain and maintain contact with the enemy.

- Provide early and accurate warnings of potential enemy attacks.
- Prevent standoff attacks.

5-22. Commanders, supported by their staffs, evaluate mission variables, focusing on the threat to establish a viable plan to control the security of the base camp. This planning includes the employment of security patrols and coordination with MSFs/TCFs.

5-23. Commanders of base camps with active airfields and landing zones consider vulnerabilities to approaching and departing aircraft and implement the necessary protection/force protection measures to counter threats including shoulder-launched and surface-to-air weapons and heavy machine guns. Ground approach and departure corridor security may also be included in a particular list of missions for the base camp. See JP 3-10 for more information.

PERIMETER ZONE

5-24. The base camp perimeter is the physical boundary of the base camp. Access to the base camp is controlled at the perimeter. The perimeter security system often forms the first significant line of defense for the base camp. This defense is accomplished through prevention, detection, and response to threat tactics. The perimeter should incorporate OPs or guard towers, fighting positions, and ECPs as applicable. Typical perimeters also include barriers in the form of wire obstacles, concrete or earthen barriers, ditches, other nonexplosive obstacles, and networked munitions emplaced to prevent or delay unauthorized access to the base camp.

5-25. Whether establishing a new base camp or occupying an existing one, the base camp commander/BOS-I initially focuses on establishing or reassessing the protective measures at the perimeter to the limit of the outer security area. Once those measures are adequate, attention is then directed to the measures used to protect personnel and critical assets on the interior of the base camp.

5-26. A properly designed perimeter security system should incorporate an integrated, layered defense in depth that takes advantage of the outer security area. See GTA 90-01-011 for more information. Commanders evaluate mission variables, focusing on the threat to establish a viable perimeter defense plan. This plan should—

- Provide adequate standoff and proper coverage of engagement areas outside of the perimeter in the outer security area. See ATP 3-90.8/MCTP 3-34B for more information.
- Limit or block all direct-fire, standoff, or ballistic-weapon sightlines from potential off-site vantage points.
- Establish fire control measures.
- Establish positive control of all personnel and vehicles entering the base camp.
- Direct the positioning and construction of OPs, guard towers, fighting positions, and ECPs in conjunction with protective positions and facility hardening to enhance survivability.
- Direct the positioning and construction of additional entry/exit points from the base camp to ensure that occupants of the base camp are not restricted to one ECP in the event they must evacuate the base camp.
- Integrate electronic security systems and unattended sensors to enhance the ability to observe and interdict potential threats throughout the base camp security area beyond the perimeter and provide early warning.
- Employ intrusion detection systems and other technology (such as biometrics, metal detectors, X-ray devices) whenever possible to improve efficiencies and effectiveness with entry and access control. See GTA 90-01-018, GTA 90-01-034, UFC 4-021-02 and UFC 4-022-01 for additional information on ECP and electronic security systems.
- Enable the defeat of a Level I threat with base camp security forces, enable the defeat of a Level II threat with the employment of a MSF, and delay Level III threats until commitment of the TCF.

5-27. Selected base camps may have designated inner and outer perimeters. Larger base camps seldom employ this double layer of perimeters, instead relying more on a single perimeter supplemented with inner barriers and access control measures around critical facilities. Creation of a double perimeter is extremely resource intensive.

5-28. Perimeter defense planning must also include a determination of adequate standoff distances from buildings and other structures that offer vantage points for the enemy outside the perimeter. Building the base camp with adequate standoff greatly enhances base camp defense. The amount of standoff needed is based on a threat assessment that considers the range, accuracy, and lethality of enemy weapons and the degree of the base camp protection measures. The protection cell and/or engineers and explosive ordnance teams determine the appropriate standoff distances to protect against blasts from IEDs and other explosive hazards. The standoff that is obtainable may be limited by the proximity of the base camp to local villages and towns and other immoveable obstructions. This is especially problematic within densely populated areas. These basic security considerations should have already been integrated into the planning, design and site selection considerations of the base camp—to include all appropriate risk mitigation. See UFC 4-010-01 for minimum standoff requirements and guidelines for new and existing construction.

INNER SECURITY AREA

5-29. This is the area inside the base camp perimeter. Interior barrier plans can be used around individual unit locations and critical assets and as traffic control measures to add depth to the base camp security plan and to halt or impede the progress of threat penetrations of the perimeter zone.

5-30. Interior security consists of those measures designed to protect personnel and assets located in the inner security area. These measures safeguard base camp critical capabilities and assets. Interior security procedures must be integrated with the overall protection/force protection plan. This ensures that security measures and systems are synchronized with a coherent protection/force protection strategy. See ATP 3-39.32 for additional discussion on interior security.

5-31. Interior security requires the integration of a wide range of protection/force protection tasks. The number of tasks, the level of execution, and the required coordination increase over time as the theater and the base camp matures. Planning considerations for internal security include base camp and HN security forces, protection of critical assets, and protection of high-risk personnel.

BASE CAMP PROTECTION FORCES

5-32. Within a joint operational area, various types of security forces are be assigned to secure the joint security area and LOCs. These include dedicated, base camp, and base cluster security forces; LOC security forces; MSFs; and TCFs. An MSF is a highly mobile, dedicated security force with the capability to defeat Level I and II threats and delay Level III threats within a joint security area. All base camp units and/or detachments must maintain a readiness posture appropriate to common attacks in the AO.

5-33. A base camp may or may not have a force dedicated to its security and defense. Therefore, it is important for base camp commanders/BOS-Is to be aware of the assets available for base camp security and defense. When there is not a dedicated security force, the base camp commander/BOS-I requires tactical control of tenant unit personnel to augment security and defense of the base camp. Base security and defense is an economy-of-force mission, so it is imperative that all organic, tenant, and transient forces residing on the base camp provide assistance to the base camp commander/BOS-I to fulfill these requirements as necessary.

ORGANIC FORCES

5-34. The base camp must be able to protect itself against a Level I threat. This is primarily accomplished with personnel and equipment assets of the unit occupying the base camp. Every Service member on the base camp should know and understand his or her assigned fighting position and responsibilities. See FM 3-90-1 and ATP 3-39.30 for more information about base camp defense.

5-35. Organic forces are typically responsible for manning perimeter fighting positions, OPs, and ECPs, constituting a standing reserve within the base camp; and conducting local security and perimeter security patrols outside the base camp, within the base camp AO. Organic forces also staff the BDOC, which integrates and synchronizes all aspects of base camp security and defense. See table 5-1.

Base Camp Security and Defense

Table 5-1. Summary of base camp protection forces

Asset	Capability	Source
Organic forces	Defend base camp from Level I threats and delay Level II threats	Unit resources; base camp commander/BOS-I forces, augmented by tenant and/or transient units
Security forces	Primarily perform security and law enforcement actions; augment base camp defense and serve as the base camp or critical facility commander's response force against Level I and Level II threat attacks.	Military police or Service equivalent
Protective services[1]	Protect high-risk personnel	Military police, CID, special agents or Service equivalent
Reinforcement forces[2]	Augment organic forces when needed	Units outside the base camp[2]
Mobile security force	Defend against and defeat Level II threats	As organized by AO commander, JFC, or JSC
Tactical combat force	Defend against and defeat Level III threats	As organized by AO commander, JFC or JSC
Allied or coalition forces	Defend against Level II and III threats	As organized by CFC
Host-nation forces	Protect and defend base camp when available to U.S. forces	Host-nation police, security, military, or other type units

Notes.
[1] Asset limited to designated high-risk personnel
[2] May include allied, coalition, or host-nation forces

Legend:
AO   area of operations
BOS-I   base operating support-integrator
CFC   coalition force commander
CID   criminal investigation division
JFC   joint force commander
JSC   joint security coordinator

5-36. A tenant or transient unit is any unit that resides on the base camp and does not fall directly under command of the base camp commander/BOS-I. While tenant and transient units typically have missions other than base camp security and defense, they are usually placed under tactical control of the base camp commander/BOS-I for base camp security and defense, and may be assigned specific security tasks while residing on the base camp. These tasks can range from security responsibilities within and around their unit areas to providing personnel to augment base camp defense and the BDOC. These tenant and transient unit requirements are coordinated by the base camp commander/BOS-I.

SECURITY FORCES

5-37. Security forces focus on the physical measures to safeguard personnel and prevent unauthorized access to facilities, equipment, and materials within the confines of the base camp perimeter. Within the Army, a military police law enforcement detachment provides police operations support to the base camp commander/BOS-I. Dedicated security forces may not be present, especially at smaller base camps. A lack of dedicated security personnel does not mean an absence of security requirements. Organic personnel may need to assume a greater role in the physical security and force protection of the base camp in the absence of security forces. Therefore it is important that the base camp commander/BOS-I and staff recognize, identify, and communicate physical security requirements early. See ATP 3-39.10 for additional details on police operations or ATP 3-39.32 for details on physical security.

Chapter 5

5-38. In the early stages of base camp operations, inner security patrols and dedicated response forces are primarily focused on security along the inside of the perimeter and on designated critical assets. As the base camp matures, especially on larger base camps with diverse units and populations, inner security transitions to a law enforcement activity with many of the same duties and responsibilities that are provided by military police units on permanent installations. On larger base camps, especially as the operational area matures, military police law enforcement detachments augmented by other military police elements on the base are typically employed to provide inner security including the provision of a dedicated response capability in the form of a military police special reaction team. This force responds, as directed, to defend critical assets and high-risk personnel within the base camp. The base camp commander/BOS-I also maintains a reserve that responds to threats to the perimeter of the base camp by repelling such attacks and finding, fixing, and defeating any perimeter penetrations by enemy forces.

PROTECTIVE SERVICES

5-39. Senior commanders and key personnel are at a significant risk of physical attacks or kidnapping for political purposes. Protective services—specially trained special agents—accomplish executive protection for designated senior commanders and other designated high-risk personnel. These special agents may be augmented by military police personnel when the situation warrants. Close coordination between in-theater assets and permanently assigned protection personnel ensures the continuity of protection. See ATP 3-39.35 for additional information on protective services.

REINFORCEMENT FORCES

5-40. By definition, the augmentation beyond the organic capabilities of base camp units are required to defeat Level II and Level III. In these situations, the base camp commander/BOS-I must coordinate for reinforcement from an MSF or TCF, depending on the level of the threat. Any unit outside the base camp that is called upon for assistance is a reinforcing force. Reinforcements are typically organized and detailed by the AO commander. The unit may be specifically dedicated to providing security and defense assistance, such as military police serving as the MSF or a dedicated TCF under control of the AO commander, a unit not currently engaged in a mission, or a combination of the two. The base camp commander/BOS-I must know from where reinforcements will originate from and the procedures for requesting assistance.

ALLIED AND COALITION FORCES

5-41. U.S. forces typically operate in a multinational coalition environments, consisting of forces from other nations, under either standing agreements such as the North Atlantic Treaty Organization (NATO), ad hoc coalitions under the United Nations, or some other authority. Responsibility for operational area security and/or base camp/base cluster security in the AO could be assigned to U.S. forces for one or more of these coalition partners, or vice versa. Coalitions require an extra emphasis on coordination, due to language barriers, differences in tactics and capabilities, rules of engagement, and other restrictions from their governments.

HOST-NATION SECURITY FORCES

5-42. HN personnel and organizations can frequently perform security functions efficiently because of their familiarity with the language, local customs, terrain, transportation and communications networks, facilities, and equipment. Much of this support may be provided by local organizations or personnel, secured through local procurement. However, HN security can be limited by the availability of resources and equipment. HN security can also be constrained by a lack of interoperability with U.S. equipment, insufficient HN capabilities, HN political issues and concerns, HN legal constraints (similar to U.S. Posse Comitatus Act restrictions), and lack of U.S. and HN agreements. When HN security assets are available, commanders should ensure that those that are dedicated to U.S. forces are used and positioned to help protect base camps. See JP 3-10 for additional details on HN security support.

BASE CAMP PROTECTION CONSIDERATIONS

5-43. The base camp commander/BOS-I applies protection/force protection considerations to protect combatants and noncombatants personnel and physical aspects of their base camps. Base camp protection considerations may include—

- **Employment of safety techniques.** Safety techniques are used to identify and assess hazards to the force and make recommendations on ways to prevent or mitigate the effects of those hazards, including fratricide avoidance. Base camp commanders/BOS-Is are the inherently responsible for analyzing the risks, implementing control measures, and to mitigating the risks. The base camp commander/BOS-I and staff factor into their analysis how their execution recommendations could adversely affect Service members. Incorporating protection within the risk management process is key. This ensures a thorough analysis of risks and implements controls to mitigate their effects. See ATP 5-19 and ADRP 3-37 for additional details on safety and risk management.
- **Implement physical security procedures.** Physical security procedures are designed to prevent unauthorized physical access to personnel, equipment, materiel, and documents on the base camp. The base camp commander/BOS-I achieves a strong physical security posture through the coordinated efforts of base camp policies, plans, and procedures. This safeguards his or her personnel and also protects against espionage, sabotage, damage, and theft. See ATP 3-39.32 for additional details on physical security.
- **Apply AT measures.** AT measures reduce the vulnerability of base camp personnel and property to terrorist acts and include limited response and containment by local military and civilian forces. AT measures emphasize detection, deterrence, and mitigation of the terrorist threat in the base camp AO. This also includes protection of HN and local civilians, as applicable. See ATP 3-37.2 for additional details on AT measures.
- **Conduct police operations.** Police operations involve policing and the associated law enforcement activities to control and protect base camp populations and resources and to facilitate the existence of a lawful and orderly environment. Police operations are conducted across the range of military operations. See ATP 3-39.10 for additional details on police operations.

BASE CAMP DEFENSE TASKS

5-44. A *defensive task* is a task conducted to defeat an enemy attack, gain time, economize forces, and develop conditions favorable for offensive or stability tasks. (ADRP 3-0) *Base defense* refers to the local military measures, both normal and emergency, required to nullify or reduce the effectiveness of enemy attacks on, or sabotage of, a base, to ensure that the maximum capacity of its facilities is available to United States forces. (JP 3-10) Successful base camp defenses share the following characteristics: disruption; flexibility; maneuver; massing effects; and operations in depth, preparation, and security. The basic defensive tasks are applicable to the area and perimeter of the base camp. See ADRP 3-90 for a discussion of these characteristics.

AREA DEFENSE

5-45. Conducting area defense is a defensive task that concentrates on denying enemy forces access to a base camp and/or surrounding terrain. The focus of the area defense is on retaining terrain where the bulk of the defending force positions itself in mutually supporting, prepared positions. Units maintain their positions and control the terrain between these positions. An area defense capitalizes on the strength inherent in a closely integrated base camp defense. The higher commander may assign subordinate units or tenant organizations the task of conducting an area defense as part of their mission. Subordinate echelons defend within their assigned AOs as part of the larger-echelon operation. See FM 3-90-1 for more information on area defense.

PERIMETER DEFENSE

5-46. The commander can employ perimeter defense as an option when conducting an area defense or in the conduct of base and base cluster defense in the echelon support. A perimeter defense is oriented in all directions. The prerequisites for a successful perimeter defense are aggressive patrolling and security operations outside the perimeter. The unit within the perimeter can perform these activities; or another force, such as the territorial defense forces of a HN, can perform them. The unit can organize a perimeter defense to accomplish a specific mission, such as protecting a fire base. A unit may also form a perimeter when it is located in the friendly echelon support area within the confines of a base or base cluster. However, divisions and corps can also organize a perimeter defense when necessary. A major characteristic of a perimeter defense is a secure inner area with most of the combat power located on the perimeter. Another characteristic is the ease of access for resupply operations. The commander coordinates direct- and indirect-fire plans to prevent accidentally engaging neighboring friendly units and noncombatants. Normally, the reserve centrally locates to react to a penetration of the perimeter at any point. See FM 3-90-1 for more information on perimeter defense.

SECURITY OPERATIONS

5-47. Base camps provide a protected location to project and sustain combat power. While some base camps, especially smaller base camps built in high-threat areas, may be required to focus on defense rather than just security, the primary focus for most base camps is not on conducting defense, except in rare instances. In these cases, normal mission operations on the base camp cease and the focus of all available assets is shifted to defense until the threat is eliminated or repelled. Once the threat is defeated, the base camp and its tenant or transient units return their focus to their primary missions.

5-48. *Security operations* are those operations undertaken by a commander to provide early and accurate warning of enemy operations, to provide the force being protected with time and maneuver space within which to react to the enemy, and to develop the situation to allow the commander to effectively use the protected force. (ADRP 3-90) The five fundamentals of security are—

- **Provide early and accurate warnings.** Early and accurate warnings of an enemy approach are essential to successful base camp protection/force protection. The base camp commander/BOS-I needs the information to shift and concentrate forces to meet and defeat the enemy. OPs, patrols, MSFs, and sensors provide long-range observation; observe enemy movement; and report the enemy's size, location, and activity.
- **Provide reaction time and maneuver space.** Base camp security assets—forces, sensors, and patrols—should work at a sufficient distance to allow the base camp commander/BOS-I time to review rapidly reported information. This timely review gives the base camp commander/BOS-I reaction time necessary to order indirect fire to slow the enemy's rate of advance; maneuver direct fire elements into place to engage, exploit, and defeat the enemy; and initiate coordination for an MSF or TCF if required, based on the threat level.
- **Orient on the force, area, or facility to be protected.** Base camp organic, security, and reinforcement forces must be aware of any enemy movement and must reposition their elements accordingly to maintain their position relative to any threats. The force must understand the base camp commander's/BOS-I's scheme of protection, including where the security force is in relation to enemy movement.
- **Perform continuous area security operations.** Base camp personnel conduct continuous area security operations to gain as much information as possible about the AO and any threats. This can be accomplished through OPs, mounted and dismounted patrols, and remote sensors such as unmanned aircraft systems (UASs) deployed to observe dead space.
- **Maintain threat contact.** Once enemy forces are detected, base camp forces must continuously collect information on the enemy's activities and disposition to assist the base camp commander/BOS-I in determining the potential and actual enemy course of action (COA) and deny the enemy the element of surprise. This requires security forces to maintain continuous visual contact, to be able to use direct and indirect fires to influence enemy actions and gain time for the base camp commander/BOS-I. Once the base camp security forces make enemy contact, they do not break contact unless the base camp commander/BOS-I or a designated security force commander specifically directs it.

5-49. Base camps typically protect their personnel and assets through the application of security activities. Some essential security activities include COMSEC, cybersecurity, information security, OPSEC, personnel security, and physical security. Of the five security tasks listed in ADRP 3-90—screen, guard, cover, area security, and local security—only area security and local security typically apply to base camps. Area security preserves the base camp commander's/BOS-I's freedom to move reserves, position fire support means, provide for mission control, conduct sustaining activities, and coordinate for reinforcing forces. Local security provides immediate protection to base camp forces and assets.

AREA SECURITY

5-50. The security of all units operating within the AO is the responsibility of the designated AO commander, who may or may not also be a base camp commander/BOS-I. This does not require that the commander conduct area security operations throughout the AO. See FM 3-90-2 for a discussion of area security responsibilities. The commander must prevent surprise and provide the amount of time necessary for all units located within the AO to effectively respond to enemy actions by employing security forces around those units. If the commander cannot or chooses not to, provide security throughout the AO, all concerned individuals and organizations are informed of when, where, and under what conditions the commander will not exercise this function. The commander generally depicts these locations using fire support coordination measures. Each base camp commander/BOS-I remains responsible for local security measures.

5-51. Area security tasks associated with base camp security employ many of the same characteristics and considerations as those documented in the defensive tasks outlined in FM 3-90-2 and ADRP 3-37. When necessary, security missions can also be used for offensive action to counter an identified threat. Commanders conduct area security tasks to provide early warnings and prevent enemy forces from combining and conducting concentrated strikes against base camps and base clusters. These operations allow commanders to provide protection/force protection to personnel and critical assets without a significant diversion of combat power.

LOCAL SECURITY

5-52. In addition to the protection and defense tasks already discussed, each base camp implements a number of local security tasks and systems. Local security consists of manning OPs, conducting local security patrols, providing perimeter security, maintaining a reserve to augment perimeter security, and performing other measures to provide close-in security for a base camp. The amount of training on these tasks and the resources devoted to conducting them in an AO depend on the mission variables. Some key base camp local security tasks and systems are—

- Rules of engagement (ROE).
- Perimeter security.
- Preplanned fires.
- Perimeter OPs and fighting positions.
- Early warning.
- Entry control.
- Patrols.
- Barriers and obstacles.
- Access control.
- Critical assets and facilities.
- High-risk personnel.
- Random operational measures.

Rules of Engagement

5-53. *Rules of engagement* are directives issued by competent military authority that delineate the circumstances and limitations under which U. S. forces will initiate and/or continue combat engagement with other forces encountered. (JP 1-04) ROE are a critically important aspect of base camp security. ROE contribute directly to mission accomplishment, enhance protection, and help ensure compliance with law and policy. Base camp commanders/BOS-Is disseminate ROE to all tenant and transient unit members, passing through their AO or base camp. Base camp security must to be planned and executed according to the standing ROE and other higher headquarters orders, which may include numerous constraints and restraints. All commanders and staff officers responsible for planning, coordinating, and executing security operations must take these factors into account. Failure to do so may have significant, possibly adverse, strategic-level consequences.

5-54. Base camp commanders/BOS-Is and their subordinates must comply with established ROE and should ensure that inconsistencies among Service components, multinational partners, and possibly even armed private security contractor personnel rules of use of force are reconciled. Discrepancies need to be resolved at the JFC's level to ensure all base camps and Services are operating with the same guidance. For more information on ROE, see Chairman of the Joint Chiefs of Staff Instruction 3121.01B. Legal considerations for ROE can be found in FM 1-04. Guidance on private security contractors can be found in DODI 3020.50.

Perimeter Security

5-55. The perimeter security system is often the first line of defense, and it provides a visual deterrent to potential adversaries. Perimeter security should be designed to incorporate the concept of layered defense in depth and integrate security elements (barriers, lighting, intrusion detection, surveillance systems, and access control equipment). Depending on the assets being protected and the resources that are available, base camp commanders/BOS-Is should provide for multiple layers of defense through the use of concentric layers of defense or barriers between the assets being protected and the perimeter.

Effective perimeter security requires a combination of physical security measures to prevent perimeter breaches. Security personnel should continuously observe and assess perimeter barriers, fencing, protective lighting and electronic security systems. See ATP 3-39.32 for more information on perimeter security. The design of the perimeter security system should—

- Provide adequate blast standoff distances. See UFC 4-010-01 and UFC 4-010-02 for more information.
- Limit or block the line of sight of operations from vantage points outside the perimeter.
- Provide sufficient room for vehicle and pedestrian access control.
- Maximize the threat ingress and egress time across the exterior site.
- Enhance the ability of security forces to observe threats before they can attack.

Preplanned Fires

5-56. Base camp commanders/BOS-Is consider all available sources of fire support based on the positioning of units throughout the operational area and coordinate with the commander assigned to the AO for obtaining that support as required. Base camp commanders/BOS-Is and fire support planners consider the benefits and risks of these systems during planning. Clearance of fires, especially in areas where noncombatants are present, can be difficult to obtain. See FM 3-09 for additional information on fire support.

5-57. When integrating fire support into the base camp defense plan, the base camp commander/BOS-I should consider several factors. These include—

- Positioning artillery or mortars to create the desired effects.
- Positioning target acquisition, counterfire, and air and missile defense early warning radar.
- Positioning air and missile defense assets, including counter rocket, artillery, and mortar intercept batteries, to deny the enemy use of airspace and destroy enemy aerial platforms.
- Establishing alternate or supplemental positions.
- Ensuring that there is sufficient ammunition on hand to support an extended engagement.
- Ensuring that target reference points are easily identifiable.
- Identifying dead space for artillery (due to the angle of fire) and covering it with mortars.
- Ensuring that all personnel, especially security force personnel, are familiar with call-for-fire procedures.
- Planning fires and likely engagement areas.
- Ensuring that all obstacles are integrated with fires.
- Registering final protective fires.

- Using specialty munitions, such as smoke and illumination rounds, to enhance the defense.
- Targeting mounted and dismounted avenues of approach, possible OP locations, and potential firing point locations.

5-58. The BDOC fire support officer is the focal point for the planning of indirect fires for base security and defense. The BDOC fire support officer coordinates with the supporting fires cell or fire support coordination center. Planned targets should include areas likely to be used as locations for standoff weapons and likely enemy avenues of approach. These targets should be planned to minimize collateral damage and civilian casualties. Copies of fire support plans and target lists must be provided to the headquarters controlling the fire support assets. Targets may be planned outside the base camp AO after coordination with the headquarters responsible for the area concerned. The BDOC and fire support coordination center ensure that all fire missions are properly coordinated to prevent the possibility of fratricide.

5-59. Fire support coordination measures and airspace coordination measures permit or restrict fires in and around base camps. Careful coordination must occur in planning these measures, especially with the HN. No-fire areas may be required to protect civilians or to prevent the disruption of missions by friendly fire.

5-60. Counterfire radar is used to determine the point of origin of indirect-fire attacks and can be used for immediate response or for pattern analysis to facilitate targeting. Air and missile defense early warning radar are used to disseminate warnings through the air defense warning system and rocket, artillery, and mortar-warn batteries. Counterfire radar is more likely to be positioned inside of base camps supporting stability tasks rather than offensive or defensive tasks.

Perimeter Observation Posts and Fighting Positions

5-61. Fighting positions and OPs provide protection/force protection while allowing enemy engagement and observation. The number of positions required on the base camp perimeter will depend on threat, security objectives, and subsequent requirements for continuous observation based on terrain effects. On mature base camps, OPs may include guard towers. The number and locations of perimeter OPs/guard towers should allow continuous observation and assessment of all physical security measures by security personnel. Effective weapon placement is also a consideration for fighting positions and OPs/guard towers. These locations should provide maximum personnel and equipment protection.

5-62. The base camp commander/BOS-I establishes mutually supportive alternate and supplementary firing positions around the perimeter, as required. Areas that are outside the perimeter and still within the base camp AO are cleared to provide good fields of fire and enable observation. Surveillance, obstacles, preplanned fires, and patrols are used to mitigate gaps and dead spaces in the defense. Engagement areas are created, as appropriate.

5-63. OPs, guard towers, and fighting positions are continually improved to increase survivability. Engineers and other elements plan and implement survivability and other protection/force protection measures. All personnel within the base camp should be assigned to cover positions and sectors in the event of an enemy attack. Otherwise, they should be directed to protective positions, or bunkers or to remain indoors. Every base camp should develop battle drills and conduct rehearsals to ensure that all residents know their role and position in the event of an attack or other hazard, such as an extreme weather event. This ensures survivability without distracting or impeding security and defense personnel. See ATP 3-37.34/MCTP 3-34C, and GTA 90-01-011 for more information on survivability operations.

5-64. OPs, guard towers, and fighting positions should be built to provide progressive levels of protection in relation to identified threats. Pre-engineered kits that can be rapidly deployed with limited manpower and equipment and can be reused are preferred. Expedient solutions that can be locally fabricated and quickly constructed by troops are also good options.

Early Warning

5-65. Early warnings are provided by ground- and aerial-based surveillance such as patrols, security cameras, UASs, unattended seismic and acoustic sensors, trip flares, and military working dogs. Civilian informants and actions of local populations are also useful indicators of pending threat actions.

Entry Control

5-66. Controlling access into and out of the base camp is critical. The objective of entry control is to prevent unauthorized personnel and vehicle access while maximizing traffic flow for authorized access. ECPs are designed and operated with a defense-in-depth approach that uses elements of distance and time to afford ECP personnel the opportunity to safely assess and react to threats. ECPs employ barriers, protective structures and other technologies to increase the protection of ECP personnel. See UFC 4-022-01 and GTA 90-01-034 for entry control design considerations.

5-67. Depending on the base camp purpose, the amount of traffic entering and exiting the base camp can be extensive. Moveable and retractable barriers and other obstacles are used to control pedestrian and vehicle traffic in and out of the base camp. Pre-engineered, easily assembled vehicle and personnel barrier designs should be used when available. These could be in kit form or locally fabricated and should be easily moved and repositioned by troops as needed.

5-68. Plan vehicle access points so that a perpendicular approach to the perimeter is minimized or eliminated. This mitigates high-speed vehicle attacks by reducing the amount of energy that can be transferred to a barrier through vehicle impact. Techniques such as sharp turns and serpentine layouts are frequently used for this purpose. See UFC 4-022-02 for additional details on vehicle barrier selection and application.

5-69. Since ECPs typically have large manpower requirements, they should be limited to the minimum amount necessary to allow the expeditious flow of traffic in and out of the base camp. The base camp commander/BOS-I can reduce the manpower requirements for ECPs by limiting operating hours for certain ECPs to peak-demand periods.

Patrols

5-70. There are two types of patrols that are relevant to base camps—internal and external. These patrols can be mounted or dismounted.

5-71. Two-person patrols are normally adequate for an internal patrol, although the exact patrol configuration is established based on an analysis of mission variables. External patrols would normally be mounted and account for the outer security area of the base camp.

Barriers and Obstacles

5-72. Barriers and obstacles are used for base camp protection through integration with observed fires, existing obstacles, and other reinforcing obstacles and defense-in-depth. Existing obstacles are inherent aspects of the terrain that impede adversary movement and maneuvers. Existing obstacles may be natural (rivers, mountains, wooded areas) or man-made (enemy explosive and nonexplosive obstacles and structures, including bridges, canals, railroads, and embankments associated with them). Reinforcing obstacles are those man-made obstacles that strengthen existing terrain to achieve a desired effect. For U.S. forces, reinforcing obstacles consist of land mines, networked munitions, demolition, and constructed obstacles. See ATP 3-90.8/MCTP 3-34B for additional information.

> *Note.* The use of some obstacles, specifically land mines, is governed by international and U.S. laws and U.S. policies. The U.S. regards land mines as lawful weapons when employed in according to accepted legal standards. The authority to employ land mines originates with the President. Since the employment of mines in foreign territories is generally considered a hostile act, the president must authorize them. Employing mines in allied territory is permissible with HN permission and Presidential authorization. U.S. forces only employ nonpersistent mines that are capable of self-destruction and/or self-deactivation. ROE and release authority define the use of land mines. See JP 3-15 for more information on the laws, agreements, and policies that are most significant for the employment of obstacles.

5-73. Barriers and obstacles are designed, constructed, and emplaced to defeat specific threats. For example, barriers used at ECPs or along high-speed avenues of approach must be capable of stopping large, high-speed, trucks. Barriers and security screens are also used to prevent casual observation on the base camp and to mitigate observation vantage points and potential sniper locations from adjacent buildings and tall structures. See ATP 3-37.34/MCTP 3-34C for more information. Additional techniques are presented in GTA 90-01-011.

5-74. Barriers, such as fencing and walls, are used to reinforce existing natural and man-made obstacles to deny or restrict unauthorized access to specified areas. Certain barriers can also enhance survivability. Barriers are available in many forms. Some are easily moveable or reusable to provide flexibility and cost savings. Barrier designs should allow for rapid construction/emplacement by troops with minimal nonspecialized tools/equipment. They should be in kit form or locally fabricated using pre-engineered designs and should if needed, be reusable. Evaluate barrier use carefully; misapplied barriers can provide a false sense of security. For instance, concrete barriers can disintegrate and become debris fragments in the immediate vicinity of large explosions. See UFC 4-022-02 and UFC 4-022-03 for details on barrier selection and application.

Access Control

5-75. Access control is used to secure all or part of the base camp from unauthorized access. In addition to the entry control measures implemented for base camp access, a personnel access control system should be established to restrict access to key facilities or areas located within the perimeter. Key facilities typically include headquarters, operations centers, and communication complexes. Generally, the unit or element owning the key facility is responsible for enforcing access control measures. In extreme situations, the base camp commander/BOS-I may designate high-value facilities or assets that require security forces to conduct security and access control. A pass and badge system, coupled with an escort system, is the most effective way to control visitors and non-CAAFs on a base camp. See UFC 4-022-01 for details on the selection of access control systems.

Critical Assets and Facilities

5-76. A critical asset is an asset that is identified as essential to the operation of the base camp and resources required for base camp protection. This includes personnel, facilities, and/or information. Critical assets are determined based on their importance in relation to the mission and may include command posts, secure storage facilities and communication nodes, ammunition, and other critical/vulnerable storage areas. For AT purposes, critical assets should also include high-population facilities (recreational activities, dining facilities, theaters, sports venues) that may not necessarily be mission-essential. Access control to critical assets requiring additional security measures should be coordinated with the unit responsible for the asset. See ATP 3-37.2 or ATP 3-39.32 for more information about asset protection and security.

High-Risk Personnel

5-77. People designated as high-risk personnel require increased security protection. Planners and security personnel must also be aware that the presence of high-risk personnel on the base camp can increase the threat to the base itself due to the increased visibility and symbolic value of a successful threat attack on the protected person(s). Risk management principles should guide the designation of high-risk personnel and high-risk billets, the approval of protective support, and determine class of the number and type of assigned protective services detail personnel. Protective services detail support is maintained at the minimal level required and employed only as necessary and appropriate based on the threat. See ATP 3-39.35 for information on protective services.

Random Operational Measures

5-78. Random operational measures are meant to change the apparent security posture of a base camp. The effect of random operational measures is difficult to measure since the rate of success is unknown, success results in fewer or no attacks. As with any changes to routine and tempo, random measures could lead to a temporary increase in accidents and equipment demands. However, assets and routines can be obscured from outside observation. Also, variations in security routines heighten the awareness of base camp personnel to security issues.

5-79. Random operational measures confuse surveillance attempts and make the base camp appear too difficult to attack. Any variation of normal routines can alter the external appearance of base camp security and frustrate enemy or adversary surveillance efforts. For example, random measures at an ECP can reduce operational predictability. Examples include—
- Move barriers and obstacles to vary traffic and serpentine patterns.
- Implement random ECP personnel shift length and changeover times.
- Change ECP operating hours.
- Use random access control procedures.
- Use random vehicle and personnel search procedures.
- Relocate ECP overwatch positions.
- Implement random roving patrols.
- Protective construction.

5-80. Protective measures should be integrated into base camp planning and construction as early as possible. In certain situations, this may not be possible. For example, initial guidance that precluded additional protective construction measures has changed. In addition, existing structures not originally considered for use on the base camp may require strengthening. Whatever the reason, protective construction implements specific measures to strengthen structures and mitigate threats.

5-81. There are two general types of protective construction. The first strengthens or reinforces structures so that they are less vulnerable to an enemy attack. This generally includes structural hardening, sidewall protection, and overhead cover. The second involves the construction of building structures specifically designed and built to protect people or assets. This generally includes bunkers, towers, and fighting positions.

5-82. There are a variety of protective construction measures available. Many are listed in ATP 3-37.34/MCTP 3-34C and GTA 90-01-011. Several standard designs are available in the AFCS. Many require significant manpower, time, and/or engineer support to emplace. Pre-engineered designs that are available as kits or easily constructed from locally available or fabricated material are preferred. They should require minimal manpower and equipment for rapid deployment by troops and should be easily upgraded to protect from more severe threats. Some of the more common protective construction measures are discussed below. Additional measures and custom solutions are available through reachback. See appendix G for more information.

5-83. Barriers are used in a variety of ways for a base camp: maintain standoff distances, establish boundaries, limit and control access of personnel and vehicles, obstruct line-of-sight observation from outside the perimeter, and protect critical assets. Barriers can be natural, man-made, or a combination of the two. Natural barriers are terrain features, such as mountains, cliffs, rivers, canals, waterways, and swamps. Man-made barriers include berms and ditches; personnel and vehicle barriers; fences; gates; and expedient barriers such as military vehicles, construction equipment, and rubble. Barriers are categorized as active (containing moving parts) or passive. Barriers are further characterized as fixed (permanently installed) or portable (moveable, although heavy equipment may be required.). Many barrier types exist; barriers should be selected based on their specific intended application.

5-84. Sidewall protection involves walls or barriers designed to stop fragments and reduce blast effects from near-miss impacts. Revetments are simple walls designed for this purpose. One of the most efficient materials for stopping fragments is a dense, granular soil such as sand. Most revetment designs are variations of techniques to hold the soil in a vertical position. Some revetment designs can also function as vehicle barriers. The primary uses of revetments on a base camp are as walls and vehicle barriers along the perimeter and at ECPs, as full-height sidewall protection for soft-sided structures such as tents and trailers, and as free-standing walls to protect mission-critical equipment. Typical sidewall protection is shown conceptually in figure 5-2.

Base Camp Security and Defense

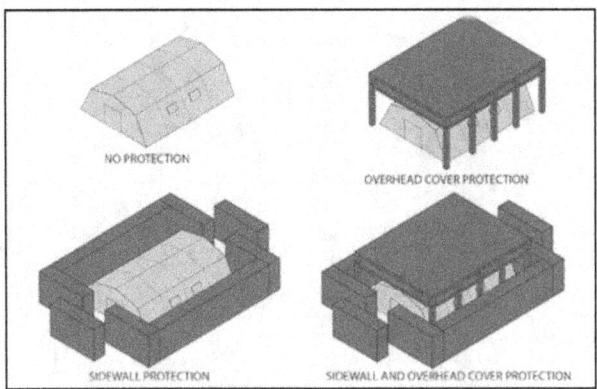

Figure 5-2. Sidewall and overhead protection

OVERHEAD COVER

5-85. Overhead cover provides protection from indirect fire and fragmentation; it also provides some protection from direct fires delivered from a higher position or from enemy aircraft. When possible, overhead cover is always constructed to enhance protection against airburst indirect-fire rounds. The basic concept is to provide a predetonation layer and shielding layer over the personnel being protected. The predetonation layer causes the fuse of the incoming round to function and detonates the round before it can penetrate. The shielding layer, located approximately 5 feet below the predetonation layer, protects against shell fragments. Overhead blast protection should always be used with adequate sidewall protection to protect from near misses.

5-86. Overhead cover dramatically increases survivability and protection. Overhead cover protection for areas with large concentrations of personnel, such as gymnasiums and dining facilities, are usually designed and constructed for a specific facility to withstand direct hits from mortars. Most individual fighting positions are not constructed to withstand a contact burst from an indirect-fire weapon.

> **DANGER**
> Improperly constructed overhead cover can collapse and result in injury or death. Eighteen inches of overhead cover provided by sandbags can weigh up to 4,000 pounds (1,814.4 kilograms) on a two-person fighting position. It is critical that positions be built according to established guidelines outlined in ATP 3-37.34/MCTP 3-34C and TM 3-34.85/MCRP 3-34.1. Do not attempt to construct a custom overhead cover solution without consulting a structural engineer.

BUNKERS

5-87. Bunkers offer excellent protection against direct-fire and indirect-fire effects. Bunkers, built above or below ground, are made of reinforced concrete, revetment material, or timber. If properly constructed with appropriate collective protection equipment, they can provide protection against chemical and biological agents. When designing a bunker, consider its purpose (command post or fighting position) and the degree of protection desired (small arms, mortars, or bombs). Prefabricated bunker assemblies (wall and roof) afford rapid construction and placement flexibility.

Chapter 5

FIGHTING POSITIONS

5-88. Fighting positions and OPs facilitate offensive and defensive operations. Aboveground positions provide an observation vantage, require less construction labor, and are easier to enter and exit than belowground positions. Conversely, these positions are harder to conceal and require large amounts of cover and revetment material. While fighting positions should provide maximum protection to personnel and equipment, effective weapon employment is the most important consideration. Weapons are often sited wherever there are natural or existing positions, and minimal digging is required.

TOWERS

5-89. Observation tower design must begin with a physical site study, including terrain analysis, and an analysis of security requirements. The location and height of the tower that best suits a particular location should be based on the nature of the base camp, the terrain under observation, the physical environment, and the functions served by the tower. Towers should be placed inside the base camp perimeter with at least a 30-foot clear zone. Towers should be located so that they are mutually supporting, with interlocking fields of view and fire. Figure 5-3 shows how increasing the distance between the tower and the perimeter wall can produce unobservable dead zones that can conceal threats. Towers should be placed so they do not create unobservable areas outside the wall, adjacent to their placement.

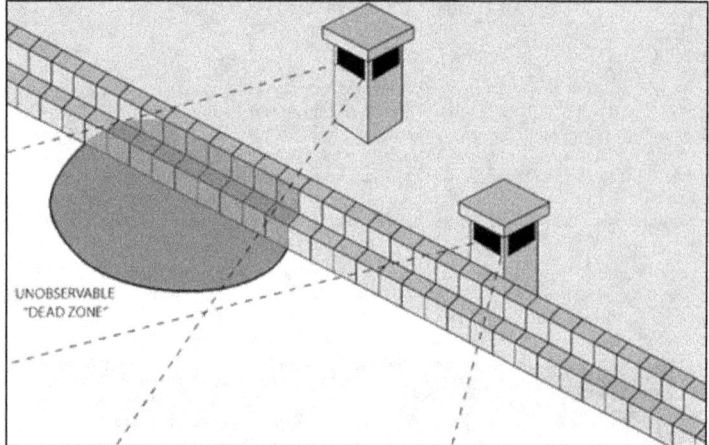

Figure 5-3. Tower location and unobservable areas

INTEGRATION OF BASE CAMP PROTECTION

5-90. All base camp protection measures are integrated and synchronized throughout the operations process. Base camp protection/force protection tasks are incorporated in a layered and redundant approach, complementing and reinforcing actions needed to achieve force protection. Base camp commanders/BOS-1s and staff plan, coordinate, and synchronize protection/force protection actions using integrating processes and continuing activities to ensure full integration of their plans.

PLANNING

5-91. Base camp defenses are planned using the processes described in FM 6-0 and MCWP 5-10. Base camp defense requirements and the tasks necessary to fulfill them are synchronized primarily through integrating processes and continuing activities. See ADRP 5-0 or MCDP 1-0 for more information. The planning process provides the framework for integrating the actions of the commander of the AO, the base camp commander/BOS-I, their staff, and others. The principles of protection are described in ADRP 3-37, and characteristics of offensive, defensive, and security tasks are described in FM 3-90-1. Additional force protection support tools are available at the USACE Reachback Engineer Data Integration (REDi) Portal and the USACE Engineer Research and Development Center (ERDC) Force Protection Portal.

5-92. The planning activity of the operations process results in a detailed base protection plan or annex. The base camp defense plan must ensure adequate protection with as small a force as necessary to avoid diminishing the ability of the base camp to function and to prevent hindering tenant units from performing their primary mission tasks. Risks to the security of the base camp are mitigated through a layered defense. A layered defense should consider the threat and design security and defense measures to protect against identified potential threats. Base camp local security preparations consider all three areas of the base camp security defense framework.

5-93. Information obtained by tenant units conducting missions off the base camp can also be critical to forming a complete threat picture and other portions of the COP. Units operating off the base camp may routinely make contact with personnel from other governmental organizations, nongovernmental organizations, local nationals, and third-country nationals. Information relevant to the overall security posture of the base may be obtained through conversation and observation with these various entities.

5-94. The application of adequate control measures is critical to a base defense plan. A control measure symbol is a graphic that is used on maps and displays to regulate forces and warfighting functions. Base defense plans use many of the same control measure used on as offensive and defensive tasks outlined in FM 3-90-1. Military symbols used to denote protection measures are found in ADRP 1-02 or MIL-STD-2525D.

5-95. Fire support coordination measures are established to permit or restrict fires in and around the base camp. No-fire areas and no-fire lines may be required to protect civilians; prevent the disruption of sustaining operations; or protect combat outposts, OPs, and patrols from friendly fire. Commanders coordinate all established controls with HN organizations to minimize interference, misunderstandings, and collateral damage. See ATP 3-09.32/MCRP 3-31.6/NTTP 3-09.2/AFFP (1) 3-2.6 for more information.

5-96. During the COA development and the array of forces (including an initial assignment of units to the planned base camp locations), planners consider the forces and equipment needed to conduct area security tasks required for base camp defense at each site, to include MSF and TCF requirements to defeat Level II and Level III threats. The required capabilities are based on initial risk assessments and the capability of available assets supporting base camp security requirements to provide early warnings and eliminate or mitigate identified threats.

5-97. The detailed information required for preparation and execution is provided in plans and orders to assist tenant and transient units with base camp defense planning. Guidelines for creating a base camp defense attachment to plans and orders are included in appendix C.

5-98. Base camp protection/force protection plans are refined based on new information from reconnaissance (including engineer reconnaissance and infrastructure assessments) and continual risk assessments. The actual location, orientation, and or site layout of the base camp and boundaries may require adjustment based on current conditions at the proposed location and the missions of subordinate units.

Chapter 5

PREPARATION

5-99. Preparation consists of activities performed by the unit before execution to improve its ability to conduct the operation. Successful base camp security and defense depends as much on preparation as they do on planning. Key preparation activities for base camp defensive tasks include—

- Planning refinement based upon IPB updates and the answering of information requirements.
- Redirecting information collection assets to focus on the most important unknowns remaining, while emphasizing the commander's critical information requirements.
- Conducting combined arms rehearsals.
- Performing precombat checks and inspections.

5-100. The rehearsal is one of the most effective synchronization tools available to commanders. Rehearsals help the staff, unit, and personnel to better understand their specific roles in upcoming operations, practice complicated tasks before execution, and ensure that equipment and weapons function properly. The execution matrix is an excellent tool to drive and focus rehearsal.

5-101. Base camp protection may involve multiple units providing elements to support the overall effort. This is especially true when employing security forces. Security forces present unique challenges since two chains of command may be simultaneously operating in one area. The possibility of confusion is always increased when multiple unit command structures are involved. The complexity associated with tasks such as clearing of fires, fire support coordination, traffic control, and communications requires close coordination.

5-102. The command of all base security forces is typically passed to the MSF or TCF commander once employed. The exact point in time must be synchronized between the base camp commander/BOS-I and MSF/TCF to ensure that the risk of fratricide is mitigated. Wargaming these actions during mission planning and rehearsals and implementing the necessary control measures and coordinating instructions are critical to avoiding fratricide.

EXECUTION

5-103. Execution involves monitoring the situation, assessing the operation, and adjusting orders as needed. Commanders continuously assess operation progress based on personal observations resulting from the direct supervision of base camp defense, information from the COP, running estimates/staff estimates, and assessments from subordinate commanders and leaders. When the situation deviates from the order, commanders direct adjustments to exploit opportunities and counter threats.

ASSESSMENT

5-104. Assessment occurs throughout the operations process, preceding and guiding the other activities, and enabling the adjustment or revision of the base defense plan and its preparation and execution. During preparation, assessment is focused on determining the friendly unit readiness to execute the base camp defense plan and an implementing any refinements to orders based on changes in the threat situation or civil considerations. During construction, initial occupation, and transitions, assessment is adjusted based on updates to assumptions and threat assessments. The focus shifts to validating the receipt posting, and understanding the COP and other SU/SA; conducting rehearsals; and that ensuring that tenant units are trained and prepared to accomplish all of their security and defense requirements and missions. It may also include ensuring that a supporting MSF/TCF is prepared to accomplish its missions in support of the base camp. The lessons that units learn while conducting or exercising base camp defensive tasks are incorporated into the base defense plan. The base camp commander/BOS-I reports lessons learned that potentially have widespread utility through the chain of command.

Chapter 6
Base Camp Transfer and Closure

Eventually, base camps are transferred to another Service, multinational force, governmental or nongovernmental organization, or the HN, or they may be closed or may be dismantled when no longer needed. Depending on any specific agreements that may exist, this process can be labor-intensive and may require the deconstruction of facilities and mitigation of environmental hazards. Additionally, there are legal and financial considerations that must be integrated to ensure that U.S. obligations are met with the least amount of cost and effort necessary.

TRANSFER AND CLOSURE PLAN

6-1. Effectively transferring or closing base camps protects U.S. interests and promotes good will and understanding with the international community. To do so, personnel must be trained in the proper procedures, doctrine and policy must be consistent, and base camp planning and design must integrate measures that facilitate subsequent transfer or closure. Likewise, Service or DOD systems must be in place to archive records including base camp plans, contractor-supplied drawings, and environmental surveys and closure reports. This includes the preservation of unit historical documentation and property, such as unit and individual memorials, as directed. To improve the efficient use of resources and preclude each unit solving the same problems, some actions may be performed by a designated unit or operational area-wide contract.

6-2. The CCDR develops the policies and procedures for base camp transfers and closures as part of the theater basing strategy. This includes guidance on the tasks to abandon, dismantle, and demolish base camps. This theater base camp transfer and closure guidance is based on operational variables; mandated timelines for force reductions, retrograde, and withdrawal as part of the exit strategy; cost-benefit analyses; existing U.S. and HN laws and regulations and agreements and negotiations with HN and private landowners. It is developed in cooperation with multinational forces and governmental and nongovernmental organizations and adjusted as developed. HN agreements are typically used to define the final condition of facilities and infrastructure on base camps being transferred to the HN, and existing buildings and land areas that were used by U.S. forces but are being returned to the HN.

6-3. The theater base camp transfer and closure guidance provides the necessary information that operational and base camp/cluster commanders will need to develop the transfer or closure plan for each base camp in their AO this plan which is developed for each base camp, details the required actions, tasks, and standards that must be completed within a certain time frame and/or within a certain sequence to ensure the base camp can be transferred or closed according to established timelines. The base camp transfer or closure plan is based on the theater base camp transfer and closure guidance, higher headquarters plans and orders, and unit SOPs. Increased management, control, and assessment may be needed during a rapid transfer or closure to ensure compliance with the policies and plans.

Note. The base camp transfer or closure plan may be part of the base camp master plan. It may also be a separate plan developed for use by the unit(s) occupying the base camp. Any plans developed for base camp transfer and closure should be maintained consistent with CCMD and Service guidance.

6-4. Commanders also establish procedures for abandoning or destroying base camps in response to an emergency or controlled evacuation. In these scenarios, sensitive items are accounted for and removed or destroyed to prevent their use by hostile forces. The base camp commander/BOS-I should establish local procedures—including evacuation routes, rallying points, and personnel accountability actions—and ensure tenant and transient units understand their requirements.

Chapter 6

GENERAL REQUIREMENTS

6-5. In addition to the routine tasks required for unit redeployments and transfers of authority, specific considerations for base camp transfers and closures are needed. These considerations can be grouped into the following five major areas:
- Environment.
- Waste disposal.
- Property.
- Contracted support.
- Records and Documentation.

ENVIRONMENTAL

6-6. The base camp commander/BOS-I is responsible for the timely identification and mitigation of negative environmental effects on the base camp. The major environmental tasks performed in support of base camp transfers and closures include—
- Conducting environmental site closure surveys.
- Transporting HAZMAT and HW to the nearest appropriate accumulation point.
- Removing AT/force protection measures to include knocking down protective berms, filling in fighting positions, and removing obstacles such as wire and vehicle barriers.
- Conducting environmental mitigation (cleaning up HAZMAT, HW, and petroleum, oil and lubricant spills).
- Disposing of medical waste and infectious wastes.
- Closing waste management facilities (solid, hazardous, medical, wastewater, and special wastes).
- Closing vehicle and aircraft washracks.
- Establishing, and later closing, equipment decontamination sites (for hazardous and biological contamination).
- Maintaining environmental documentation.

Note. See ATP 3-34.5/MCRP 3-40B.2 for more information on environmental considerations.

6-7. Based on the theater guidance for transfers and closures, and in coordination with higher headquarters, base camp commanders/BOS-Is develop specific procedures and assign tasks. This ensures the—
- Availability of proper personnel, supplies, and equipment to properly package and ship hazardous and special waste (such as approved containers, labels, placards, and material safety data sheets).
- Availability of proper personnel, supplies, and equipment to clean up any identified or anticipated area that will likely require action.
- Completion of necessary environmental documentation.
- Proper decontamination and packing of unit equipment and the proper disposal of decontamination waste. Guidelines are available in ATP 3-11.37, UFC 1-201-01, and UFC 3-190-06. Additional techniques are presented in GTA 05-08-016.
- Proper transport and turn-in of HAZMAT and HW to the designated accumulation point.
- Waste disposal.

6-8. Based on theater guidance, the base camp commander/BOS-I develops a plan for disposing of all remaining waste on the base camp and returning each waste collection, accumulation, and treatment site to its preexisting state or the required condition for closure or transfer. Unit or base camp environmental officers should make initial contact with the appropriate environmental officer or designated representative for removal of hazardous and special waste from HW accumulation points at least 60 days in advance of the transfer or closure date. The plan should address the following critical areas:
- Disposition of reusable and recyclable materials.
- Requirements for properly packaging, inventorying, labeling, and turning-in hazardous and special waste for disposal and cleaning up of HW accumulation areas.

- Termination of waste management contracts, removal of contractor-furnished equipment, and clean-up of the surrounding area.
- Disposition of empty hazardous and special waste containers, to include standards for turn-in.
- Removal of fuel bladders, blivets, secondary containment liners, and associated fuel distribution equipment and establishment of clean-up standards necessary for any affected areas.
- Disposition of secondary containment and protective berms.
- Disposition of waste material generated from base camp deconstruction.
- Closure and clean-up of all waste management areas, such as incinerators, landfills, recycling operations, composting sites, and land farming operations.
- Disposition of medical waste.
- Proper shutdown of water purification systems, the disposition of the wastewater and brine lagoon, and the need for a water survey.
- Disposition of wastewater treatment systems.
- Disposition of above ground and underground storage tanks.

WASTE DISPOSAL

6-9. When closing base camps, all latrines, soakage pits, landfills, trash burial sites, and septic systems must be closed, and marked, and their locations recorded and archived for future reference. While simple methods will generally involve only covering with earth, agreements with the HN may require more detailed methods and some form of long-term monitoring to detect potential groundwater contamination. In the absence of formal guidance, best management practices must be used. This may require consultation with environmental experts to ensure the best possible solutions. See TM 3-34.56/MCIP 3-40G.2i and ATP 3-34.40/MCTP 3-40D for more information.

6-10. Units record the grid coordinates and take post-closure digital photographs of each waste management site. This information is incorporated into the environmental site closure report and is archived for future reference.

PROPERTY

6-11. Ensuring the proper disposition (retain, reutilize or redistribute, retrograde, or dispose) of property is critical to transfers and closures. Commanders at all levels share in the responsibility for implementing the controls necessary to that ensure accurate and complete official records are maintained for all property transfers. The types of property to be addressed during transfers and closures include—

- **Real property.** Real property refers to land and permanent improvements to that land. This includes structures, buildings, and equipment such as heating systems, installed carpeting, and overhead hoists affixed and built into the facility as an integral part of the facility. Economic improvement of the HN is considered when recycling or transferring facility infrastructure. See AR 735-5 for more information.
- **Personal property.** This is any property that can be moved and reused without significant refurbishment or degradation from its intended purpose. Personal property includes government property and items owned by individuals. Some personal property items, such as lighting, plumbing, and environmental control units, become real property once installed within buildings or facilities. See AR 735-5 for more information.
- **Contractor-managed, government-owned property.** This includes government-furnished equipment and contractor property acquired with government funds.

6-12. There may be HN government and private owners for various land parcels located inside the base camp footprint. A critical task includes identifying the rightful landowner so that the necessary negotiations and lease payments can be made. Depending on the viability of the HN government and the availability of land records, U.S. forces may need to effect deed verification to facilitate the timely disposal of real estate. USACE, NAVFAC, and the Air Force Civil Engineer Center (AFCEC) have experts who can deploy or provide reachback in support of these requirements. See EP 500-1-2 for more information.

6-13. In preparation for transfers and closures, base camp commanders/BOS-Is and tenant unit commanders begin by conducting property inventories and identifying excess property. Excess property is property that is not be included in the base camp transfer and is not contractor-owned or part of a unit's modified table of organization and equipment. Serviceable or reparable excess property may be redistributed or cross-leveled to other camps to fill shortages or turned in to a Defense Logistics Agency disposition services facility. Unrepairable and nonrecoverable excess property is disposed of through recycling or an approved waste disposal facility, as directed. Some items may require demilitarization or destruction prior to transfer or disposal to prevent reuse or exploitation. Planners ensure that there is adequate and proper space for property being redistributed to other base camps and that the necessary transportation requirements are coordinated. Base camp commanders/BOS-Is and tenant unit commanders should also work with their supporting contracting element and/or LOGCAP forward personnel to ensure that the supporting contractors are prepared to transport or properly dispose of any contractor-owned equipment.

6-14. Land areas may need to be restored to a certain condition based on HN agreements and negotiations with landowners. This may include removing gravel surfaces, concrete pads and footings, survivability measures (such as protective berms and fighting positions), and AT/force protection measures (such as concrete barriers and wire obstacles) unless the HN requests that those devices be left in place for continued use. Areas used as ranges and ammunition supply points may require clearance actions to remove or mitigate explosive hazards.

CONTRACTED SUPPORT

6-15. Contracted support is often an integral part of base camp operations. Base camp commanders/BOS-Is determine which contracts to retain in order to sustain essential base camp support and services. Other contracts that are unnecessary can be revised or closed out to reduce costs and ensure that transfers and closures stay on track. They begin by identifying all open contracts, including ongoing material requisitions. They also review construction contracts to determine those that should continue to move forward and those that should be terminated based on cost-benefit analyses and the planned transfer or closure date of the base camp.

6-16. Base camp service and support contract requirements should be reduced, often referred to as descoped or adjusted, as the base camp population decreases. The base camp commander/BOS-I coordinates with owners of required resources and determines the contracted services and support that are mission-essential or needed for life, health, and safety.

6-17. Commanders of remaining base camps should ensure that their base camp support and services contracts are modified to accommodate the population expansion resulting from realignment and consolidation. Contractors and vendors must be given adequate advance notice of closures so that they can plan and execute their recovery and/or redeployment plan. Commanders ensure accountability of contractors as contracts are closed and base camps are transferred and closed so that unauthorized personnel do not remain on base camps.

RECORDS AND DOCUMENTATION

6-18. Commanders maintain and archive base camp records and documents to provide a historical record that facilitates base camp transfers and closures and the development of lessons learned. This establishes the baseline for that location in the event that follow-on operations or legal actions are required. The CCDR confirms standards of master plans and contractor-provided documentation and archives in a central Service or DOD repository. Critical base camp documents that are maintained, transferred to incoming commanders, and archived upon closure include the following:

- Base camp master plans (including contractor-provided drawings).
- Environmental documentation, to include—
    - DD Form 2993, *Environmental Baseline Survey (EBS) Checklist*.
    - DD Form 2994, *Environmental Baseline Survey (EBS) Report*.
    - DD Form 2995, *Environmental Site Closure Survey*.

- Environmental closure reports. See the ATP 3-34.5/MCRP 3-40B.2 for more information.
- Sampling reports, to include chain-of-custody forms and analysis results.
- Spill reports.
- Results of environmental inspections.
- Waste turn-in documents and removal manifests for HW.
- Documentation of solid waste disposal into burn pits, incinerators, and landfills.
- Environmental site closure reports generated based on the environmental site closure survey.
- Documentation of all clean-up actions taken on-site.
- Real estate documents, to include—
  - Deed verification map and deeds in English and the predominant language for the AO/HN.
  - Leases and other contractual documents related to real estate acquisition.
- Support contract records.
- Preliminary excess personal property inventory.
- Approved property change notification.
- Property inventories.
- Property disposition documents.
- Final closure or transfer documents.
- Legal reviews.

6-19. Accurate records facilitate the transfer of base camps by providing the new base camp commander/BOS-I with detailed information on building plans, infrastructure locations, and environmental considerations. These records assist in maintaining the base camp and in preventing or mitigating hazards. Base camp records are also essential to base camp closure by providing information on base camp infrastructure that is to be dismantled, assisting in the planning process, helping to mitigate safety and environmental issues, and providing baseline information that helps protect the U.S. government against potential liability claims. In addition to facilitating closure or transfer actions, maintaining base camp archives provides information that can assist base camp planners in the future by providing planning and operational information

This page intentionally left blank.

Appendix A
Metric Conversion Chart

This appendix complies with AR 25-30, which states that weights, distances, quantities, and measurements contained in Army publications will be expressed in U.S. standard and metric units. Table A-1 is a metric conversion chart for the measurements used in this manual. For a complete listing of preferred metric units for general use, see Federal Standard 376B.

Table A-1. Metric conversion chart

United States Units	Multiplied By	Equals Metric Units
Feet	0.3048	Meters
Inches	0.0254	Meters
Metric Units	**Multiplied By**	**Equals United States Units**
Meters	3.2808	Feet
Meters	39.3700	Inches

This page intentionally left blank.

Appendix B
Base Camp Master Planning and the Military Decision-Making Process/Marine Corps Planning Process

Base camp master planning provides the foundation for a successful, operational base camp. This appendix illustrates the similarities and differences in the MDMP/MCPP within base camp master planning.

For Marine Corps users: Appendix B references Army annexes, appendices, and tabs that do not align with Marine Corps annexes, appendices and tabs. See MCWP 5-10 for correct Marine Corps annexes, appendices, and tabs.

CONSIDERATIONS WITHIN THE PLANNING PROCESS

B-1. The *MDMP* is an iterative planning methodology used to understand the situation and mission, develop a courses of action, and produce an operation plan or order (ADP 5-0). The MCPP is a methodology, that helps organize the thought processes of the commander and staff throughout the planning and execution of military operations (MCWP 5-10). The processes are similar to the joint operation planning process (JOPP), although the steps are slightly different. See JP 5-0 for more information.

B-2. The MDMP/MCPP helps leaders apply thoroughness, clarity, sound judgment, logic, and professional knowledge to understand situations, develop options to solve problems, and reach decisions. The process helps commanders, staffs, and others think critically and creatively while planning. The end result of the MDMP/MCPP is an OPORD or OPLAN with a clear, concise concept of operations and required supporting information. The MDMP is detailed in FM 6-0; the MCPP is detailed in MCWP 5-10.

B-3. In simple terms, base camp development planning, is master planning focused on base camps. It is accomplished in a manner similar to that for any system or decision that requires a coordinated and synchronized set of steps or actions to accomplish a long-term vision and subsequent objective. Master planning facilitates this with a set of steps similar to the MDMP steps. Since the MDMP/MCPP is the primary planning tool used by the military, it is reasonable that it would be the preferred method for base camp planning. It is important to understand that whether the steps of the MDMP/MCPP or master planning are used to reach decisions about base camp development planning, the results will be the same. In a general sense, the steps of the MDMP or the master planning process provide the methodology to collect, organize, and evaluate data.

B-4. During the MDMP/MCPP, the staff identifies base camp requirements and critical items and services needed to project and sustain combat power. Table B-1, page B-2, shows the similarities between the steps of the MDMP/MCPP and base camp planning and highlights some of the key base camp considerations in relation to the following planning steps:
- Receipt of mission/problem framing.
- Mission analysis/problem framing.
- COA development.
- COA analysis/COA wargaming.
- COA comparison and decision.
- COA approval.
- Orders production, dissemination, and transition.

Appendix B

Table B-1. Base camp planning considerations during the planning process

Steps of the MDMP	Steps of the MCPP	Base Camp Planning Considerations
Receipt of the mission	Problem framing	<ul><li>Identify potential sources of data and information to include existing assessment products such as EBSs, OEHSA, and infrastructure assessments.</li><li>Request geospatial information and terrain visualization products to help understand terrain effects.</li><li>Request intelligence products on potential threats to the base camp.</li><li>Gather information on the local population to determine its effect on possible base camp locations.</li><li>Update running estimates/staff estimates.</li><li>Disseminate base camp-relevant information to the appropriate staff sections for inclusion in their running estimates/staff estimates.</li></ul>
Mission analysis	Problem framing (continued)	<ul><li>Understand the higher command's basing strategy.</li><li>Assess assets available to perform base camp life cycle activities (joint and multinational forces, host nation, and contractors), identify obvious shortfalls, and prepare requests for augmentation for the commander's approval.</li><li>Determine constraints to include—<ul><li>Allowable design and construction standards in theater-specific guidelines.</li><li>Higher headquarters policies, procedures, plans, orders, and directives.</li><li>Joint and Army/Marine Corps directives and regulations.</li><li>International and U.S. laws and regulations as applicable.</li><li>Host-nation laws and local customs and practices.</li></ul></li><li>As part of the initial intelligence preparation of the battlefield/battlespace—<ul><li>Evaluate terrain and weather effects on base camp activities.</li><li>Evaluate the effects of adversaries and neutrals on base camp activities.</li><li>Assess the availability of existing facilities and infrastructure within the operational area, and develop facts and assumptions to support assessments.</li><li>Identify potential base camp locations based on threat patterns and terrain.</li></ul></li><li>Identify specified and implied base camp tasks and recommended essential base camp tasks, determine any obvious shortfalls in assets available, and initiate requests for support or augmentation as early during planning as possible.</li><li>Integrate information requirements and engineer or other necessary specialized reconnaissance capabilities into the information collection plan.</li></ul>

Table B-1. Base camp planning considerations during the planning process (continued)

Steps of the MDMP	Steps of the MCPP	Base Camp Planning Considerations
COA development	COA development	<ul><li>Integrate the base camp principles.</li><li>Refine base camp requirements and possible solutions based on mission variables.</li><li>Recommend base camp locations based on the—<ul><li>Availability of existing facilities and infrastructure.</li><li>Terrain, environmental, and civil considerations.</li><li>Threats to base camps.</li><li>Ability to sustain and secure base camps in a specific area.</li></ul></li><li>Allocate base camp capabilities based on identified requirements (troop-to-task analysis).</li><li>Identify nodes and linkages of base camps, including the formation of base clusters.</li></ul>
COA analysis	COA wargaming	<ul><li>Identify advantages and disadvantages of base camp design solutions using the following evaluation criteria developed before wargaming, such as—<ul><li>**Protect.** The ability to employ response forces and first responders in response to attacks and emergencies.</li><li>**Sustain.** The ability to access base camps for services, resupply, and casualty evacuation.</li><li>**Maneuver.** Mission support to maneuver units</li></ul></li><li>Wargame (action/reaction) enemy attacks and emergencies on base camps and the employ response forces and first responders.</li></ul>
COA comparison	COA comparison and decision	<ul><li>Analyze and evaluate advantages and disadvantages of each COA from a base camp perspective using the evaluation criteria developed before wargaming.</li></ul>
COA approval		<ul><li>Gain approval for any changes to the essential tasks for base camps.</li><li>Gain approval for recommended priorities of effort and support.</li><li>Gain approval for requests for base camp augmentation to be sent to higher headquarters.</li><li>Initiate real estate acquisition actions once base camp locations have been approved.</li><li>Provide commander with updates on base camp issues or concerns within the COA decision briefing as appropriate.</li></ul>
Orders production, dissemination, and transition	Orders development / Transition	<ul><li>Integrate base camp tasks within the plan or order, and produce the base camp appendix.</li><li>Ensure the quality and completeness of the subordinate unit's instructions for performing base camp life cycle tasks.</li></ul>

Note. The Army uses the MDMP, and the Marine Corps uses the MCPP. The processes are similar, although the steps are different. The MDMP is described in FM 6-0; the MCPP is described in MCWP 5-10.

Legend:
COA   course of action
EBS   environmental baseline survey
FM   field manual
MCPP   Marine Corps planning process
MCWP   Marine Corps warfighting publication
MDMP   military decision-making process
OEHSA   occupational and environmental health site assessment
U.S.   United States

Appendix B

RECEIPT OF MISSION/PROBLEM FRAMING

B-5. During receipt of mission/problem framing, the staff gathers existing information that will facilitate base camp planning. The information needed is obtained from multiple sources, to include: self-discovery through the planning process, another unit's or organization's assessments, studies or reports of recent operations or activities in the area, higher headquarters basing strategy, a supporting unit such as a forward engineer support team or engineer facilities detachment, and nondeployed supporting organizations such as USACE base camp development teams accessed through reachback. The staff focuses on gathering information such as—

- Command policies and directives on facility allowances and construction standards (base camp standards) that apply to the operational area or region.
- Existing geospatial information and terrain visualization products (see ATP 3-34.80) that help indicate where it may be best to develop base camps based on—
    - **Accessibility**—by ground and air (proximity to established LOCs).
    - **Vulnerability and defensibility**—based on enemy and terrain considerations such as those expressed in OAKOC/KOCOA.
    - **Constructability**—based on soil composition, surface and subsurface configuration, slope, access to construction materials and services, and the availability of existing facilities and infrastructure.
    - **Suitability**—based on civil and environmental considerations such as proximity to cultural, religious, and historical sites; environmentally sensitive areas; and areas that affect the local population.
- Existing intelligence products on potential threats to base camps.
- Country studies and information on local populations and economies to include—
    - Existing facilities and infrastructure.
    - Local markets for potential sources of labor, services (such as waste disposal and recycling), and materials that could support base camps.
- Existing infrastructure assessment or reports, EBSs, or OEHSAs.

MISSION ANALYSIS/PROBLEM FRAMING

B-6. Determining base camp requirements and developing preliminary estimates for material and construction requirements begin during mission analysis/problem framing. These requirements and preliminary estimates are further developed and refined as planning progresses and are updated and completed in more detail during design. JCMS is helpful in providing planners with rough estimates on resources and time needed to build standard facilities. As part of mission analysis/problem framing, the staff analyzes the higher headquarters plan or order to understand the higher headquarters concept or strategy for base camps within the operational area and timelines. Most of the details pertaining to base camps are found within the protection, sustainment, and engineer annexes.

B-7. As part of IPB, which begins during mission analysis/problem framing and continues throughout execution, the staff focuses on better understanding the effects of the terrain, adversaries, and environmental and civil considerations on specified base camp locations or on determining potential base camp locations. Coordination is conducted with the assistant chief of staff, intelligence/intelligence staff officer, for geospatial engineering support in analyzing the terrain and generating geospatial information and the corresponding terrain-visualization products to facilitate SU/SA. See ATP 3-34.80 for information on geospatial engineering.

Determining Possible Base Camp Locations (Site Selection)

B-8. The staff determines possible locations for base camps based on an analysis of operational and mission variables, with added emphasis on terrain, civil, and environmental considerations. The operational commander may wish to locate a base camp to best support the projection of combat power; cost, sustainability, real estate acquisition, or protection considerations may support different locations. See base camp site selection considerations in relation to the mission variables in table B-2.

Table B-2. Site selection considerations in relation to mission variables (METT-TC/METT-T)

Mission Variables	Site Selection Considerations

Mission	• Analyze the unit mission to determine the purpose of base camps and the major functions they must perform based on tenant and transient unit operational requirements, to include—
	  ▪ Requirements for specific types of facilities such as airfields, landing zones, ammunition supply points, and firing ranges.
	  ▪ Types and sizes of tenant units (land area requirements).
	  ▪ Future requirements (sufficient land area for expansion; accessibility; and access to sources of water, power, and energy).

Enemy	• Analyze threats to the base camp and the associated protection considerations such as proximity to populations, standoff, and perimeter requirements.
OAKOC/KOCOA	• Vegetation—effects on movement, landing zones, observation, and cover and concealment. • Hydrology—access to water and avoidance of surface drainage. • Soil composition—suitable for construction, trafficability, and waste management options. • Surface and subsurface configuration—trafficability; cut, fill, and clearing requirements; natural slope for drainage; seismic conditions; and clear line-of-sight for communication and collection systems. • Obstacles—natural and man-made impediments (including the presence of people) to base camp construction, operations, and sustainment. • Man-made features—existing structures and local facilities and infrastructure that affect base camps.
Troops and support available	• Availability of local workers, equipment, and services to perform base camp construction and operations tasks.
Time available	• Time available for construction (based on when the constructing unit can occupy the site and the delivery of construction materials and equipment).

ASCOPE	• Relationship with the local population (acceptance and tolerance).
	• Local political climate and perceptions and the effects on location, design, and land use decisions. Politically unpopular decisions may attract acts of aggression.
	• Effects traffic, explosive safety, inconvenience on adjacent landowners.
	• Proximity to historical, cultural, religious, and environmentally sensitive areas.
	• Areas to include—
	  ▪ Sources of natural construction resources (water, gravel, fill materials).
	  ▪ Political, ethnic, or tribal boundaries and locations of government centers.
	  ▪ Structures—availability of existing structures and local facilities and infrastructure that can help sustain base camps.
	  ▪ Capabilities—ability of local economies and local businesses and laborers to support base camps.
	• Organizations within and outside of the AO that can support or affect base camps. This includes—
	  ▪ Local labor unions.
	  ▪ Criminal organizations.
	  ▪ Community watch groups.
	  ▪ Governmental and nongovernmental agencies and organizations.
	• Effects of indigenous and transient civilians on base camps (dislocated civilians).
	• Routine, cyclical, planned, or spontaneous activities that can affect base camps (holidays, elections, celebrations, demonstrations).

Legend: AO area of operations ASCOPE area, structures, capabilities, organizations, people, and events KOCOA key terrain, observation and fields of fire, cover and concealment, obstacles, and avenues of approach OAKOC observation and fields of fire, avenues of approach, key terrain, obstacles, and cover and concealment	

B-9. Site selection begins during mission analysis/problem framing with the identification of suitable and unsuitable areas that aims to narrow options and facilitate timely COA development. These areas are primarily determined based on an analysis of terrain and civil considerations. An example of a suitable area

Appendix B

is an area with adequate existing facilities and infrastructure or readily accessible construction resources such as materials and labor pools. Unsuitable areas, which should generally be avoided, include areas such as those that are prone to flooding, have severe slopes or dense vegetation, or are inaccessible to heavy construction equipment and areas that are environmentally sensitive or that have historical, cultural, or religious significance. Tailored geospatial products can be developed to show suitable and unsuitable areas to help visualize the terrain. Site selection refinement continues throughout the planning and preparation phases based on the results of information collection efforts.

B-10. Real estate acquisition is a key task in support of site selection. Right of ways or easements may also be required for transportation and utility distribution lines. CCDRs are responsible for the coordination of real estate requirements within their AORs. USACE contingency real estate support teams, NAVFAC, and AFCEC have experts who can deploy or provide reachback in support of these requirements. The contingency real estate support team is a deployable team that can support any echelon, it is typically tailored to support an Army component headquarters configuration with support missions requiring real estate management. This team operates as augmentation to the supported force engineer staff or supporting engineer headquarters. See EP 500-1-2 for more information.

Terrain Considerations

B-11. Base camps are ideally located in terrain that is defensible, suitable, and sustainable. Terrain considerations are further described as follows:
- **Defensible**—based on terrain effects on specific equipment, weapons, and employment methods and the vulnerability of base camp occupants and critical infrastructure based on enemy observation and fields of fire. In general, planners avoid locations that are adjacent to higher surrounding terrain or buildings that provide easy observation (vantage points) onto the base camp. Base camps are best situated in areas where there is a 360-degree unobstructed view around the camp exists or such a view can be established by clearing.
- **Suitable**—based on construction considerations pertaining to soil composition, hydrology, and surface/subsurface configuration; elevation analysis for positioning line of sight-based communication and collection systems and environmental considerations; and the religious, cultural, and historical significance of an area.
- **Sustainable**—based on accessibility by air and/or ground LOC, and proximity and access to existing sources of water, power, energy, and construction sources and materials.

B-12. In situations where existing facilities are used or base camp locations have been specified, planners must still analyze the effects of terrain. They can then determine ways to mitigate any aspects of the terrain that are unfavorable for base camps.

B-13. Planners analyze the natural and man-made features in an area and evaluate their effect on base camps as part of IPB. They address the six characteristics of terrain using the five military aspects of terrain expressed in the OAKOC/KOCOA. Terrain considerations in relation to OAKOC/KOCOA are shown in table B-3, page B-8. The six terrain characteristics are further explained as follows:
- **Vegetation.** Vegetation includes trees, scrubs and shrubs, grasses, and crops (cultivated areas). Base camp planners, aided by geospatial engineers/geographic intelligence specialists, analyze the effects of vegetation on vehicular and foot movements, landing zones, drop zones, observation, and cover and concealment.
- **Hydrology.** Water is an essential commodity and is always an important factor in planning base camps. It is necessary for drinking, sanitation, food preparation, and construction. Certain support activities, such as helicopter maintenance and the operation of medical treatment facilities, consume large volumes of water. Because of the importance of water, planners identify alternate sources and backup means for producing, treating, and distributing it. Planners should always consider the effects of base camp water usage on the local population, economy, and agriculture. Untreated or stagnant water can present health hazards. Through terrain analysis, geospatial engineers/geographic intelligence specialists can help planners determine probable sources of water that may exist on and below the surface. Surface drainage, such as streams, rivers, wet or dry watercourses, and areas prone to flooding or flash flooding, can affect accessibility to base camps and render low-lying land areas unusable. Base camp planners must consider the flow and

channeling characteristics of surface water which vary based on geographic location and seasonal weather patterns. Planners must also consider proximity to dams, levees, and other drainage features that could result in catastrophic effects if they fail. Planners should contact the water resources team at the Army Geospatial Center for additional support.

- **Soil composition.** Soil composition includes soil type, drainage characteristics, and moisture content. Soil composition can affect trafficability, road and airfield construction, waste management options, and the ease of digging fighting positions in a specific area. Precipitation is an important factor to consider since it can change the characteristics of soil. Generating soil data normally requires extensive field sampling and the expertise of soil analysts. Once the data is acquired, geospatial engineers/geographic intelligence specialists use it in combination with standard geospatial products and imagery to create tailored geospatial products that enable further staff analysis. See TM 3-34.64 and ATP 3-34.80 for more information.
- **Surface and subsurface configuration.** Surface and subsurface configuration refers to the physical shape of the terrain and includes elevation, slope, surface roughness, and seismic conditions. Slope and the local relief, which is the difference in elevation between points in a given area, affect trafficability, construction requirements (cut and fill requirements), and structural designs such as gravity-fed water utilities. Surface roughness can include uneven surfaces, jagged rocks, and debris that can affect such things as aircraft landings, vehicle movements, and the positioning of prefabricated structures and buildings. Seismic hazards within an area, based on seismic zone, soil conditions, and structure use, may affect base camp locations and the seismic design of structures.
- **Obstacles.** Obstacles (natural or man-made) are any impediments that affect the construction, functioning, or sustainment of base camps. Examples of natural obstacles include rivers, forests, mountains, and steep slopes. Examples of man-made obstacles include buildings, structures, explosive hazards, and the presence of civilians. Some terrain or specific areas may present an obstacle based on religious, political, historical, or environmental significance. See cultural obstacles described in ATP 3-90.4/MCTP 3-34A.
- **Man-made features.** These include existing structures, facilities, and infrastructure that can positively or negatively affect base camp development. Existing structures, facilities, and infrastructure—to include underground utilities—may reduce requirements for new construction if they do not impose any health or environmental hazards that cannot be mitigated.

Weather Considerations

B-14. Planners consider the effects of weather on the design and performance of base camp facilities and infrastructure. Weather information is normally prepared by staff weather officers and distributed through intelligence channels. The primary weather conditions that planners should consider with respect to base camps are—

- **Temperature and humidity.** Extreme temperatures can affect construction efforts and the efficiency and effectiveness of base camp facilities and infrastructure. Extreme cold can impede digging, and freezing water affects the flow of water through piping systems.
- **Precipitation.** Rain and snow can affect road trafficability and the ability to transport materials and supplies. Heavy rainfall and snowmelt can render low-lying areas unusable or cause mudslides, and storm water runoff can cause containment systems to overflow and contaminate surrounding areas.
- **Wind.** Acknowledge of the prevailing wind direction is important for positioning base camps upwind from local agricultural, industrial, and waste areas and for positioning base camp waste management systems downwind from troop billeting, work areas, and airfields to reduce the effects of odors and toxic smoke and fumes. Wind speed is also an important consideration in determining the feasibility of wind as a source of energy. Wind and its behavior in low-lying areas should also be considered in vulnerability assessments regarding hostile airborne contaminant attacks.

Appendix B

Table B-3. Terrain considerations in relation to OAKOC/KOCOA

Military Aspects of Terrain	Terrain Considerations for Base Camps
Observation and fields of fire	• Effects of natural and man-made features (trees, fences, buildings) on electronic and line-of-sight surveillance systems, and unaided visual observation. • Defensibility of the area based on terrain effects on the trajectory of munitions (direct and indirect fire) and tube elevation. • Vulnerability of the base camp based on enemy observation and fields of fire (vantage points for direct-line-of-sight weapons).
Avenue of approach	• Base camp accessibility based on air and ground avenues of approach. • Avoidance of sites that are close to main thoroughfares with uncontrollable or straight-line vehicular access.

Key terrain	• Nominations for key terrain are based on the mission, concept of operations, threat, environment, and civil considerations.
	• Examples of key terrain considerations based on the environment include—
	  ▪ Complex terrain: tall structures, choke points, intersections, bridges, and industrial complexes.
	  ▪ Open environment: terrain features that dominate an area with good observations and fields of fire, choke points, and bridges.

Obstacles	• Effects of natural and man-made obstacles on construction, sustainment, protection, communications, and other base camp tasks. See framework of obstacles in ATP 3-90.4/MCTP 3-34A.
	• Impediments to ground movements include—
	  ▪ Slope.
	  ▪ Vegetation.
	  ▪ Road characteristics (curves, slope, width, clearance, and load bearing [bridge classification]).
	• Impediments to air movements include—
	  ▪ Elevations that exceed aircraft service ceilings.
	  ▪ Vertical obstructions such as buildings, power lines, and communication towers.
	  ▪ Bird habitats or attractions (such as waste sites, marshes, and wetlands) that can increase the risk of bird strikes.
	• Other obstacles to consider include—
	  ▪ Structures (such as dams, industrial chemical plants, and other hazardous sites) that are potentially hazardous if damaged or destroyed by a force of nature or act of man.
	  ▪ Cultural obstacles such as religious tracts of land, historical sites, and environmentally sensitive areas.
	  ▪ Human obstacles such as crowds and vehicle traffic.

Cover and concealment	• Aspects of the terrain that offer protection from bullets, exploding rounds, and explosive hazards (cover). • Aspects of the terrain that offer protection from observation (from aerial and ground detection), such as vegetation and surface configuration (concealment).

Legend:
ATP   Army techniques publication
MCTP Marine Corps tactical publication

Civil Considerations

B-15. Civil considerations help commanders understand the social, political, and cultural variables within the AO and their effect on base camps. The staff analyzes civil considerations in terms of the categories expressed in the following memory aid: areas, structures, capabilities, organizations, people, and events. See ADRP 6-0 for more information.

Environmental Considerations

B-16. The operation of base camps and other installations, such as airfields, ports, and detention facilities of war camps, requires the integration of environmental considerations. This begins early in the planning phase (with special emphasis during site selection) and continues throughout the base camp life cycle. See ATP 3-34.5/MCRP 3-40B.2 for more information.

B-17. The existing infrastructure and the surrounding area are surveyed to help planners determine the best location for a base camp from an environmental and health perspectives. This survey requires personnel with the necessary expertise to identify potential hazards and may require samplings of the air, soil, and water. Gaining insight about previous site use on and around the area is helpful in determining potential hazards. Factors such as evidence of environmental contamination, landfills or trash burial sites, and surrounding land uses are considered.

B-18. The EBS and the OEHSA are an important part of base camp development and must be conducted for every site that is occupied by U.S. forces. An EBS and OEHSA should be conducted as early as possible in the planning and design phase to allow for any mitigation or adjustments and ensure there is no wasted construction effort. The EBS and the OEHSA are assessment tools to identify potential health hazards and environmental contamination. EBS and OEHSA automated systems should be linked to a common database that archives results and populates reports. See the USAES Environmental Baseline Survey, OSHA Field Safety and Health Manual, and ATP 3-34.5/MCRP 3-40B.2 for more information.

B-19. An OEHSA is conducted to determine whether environmental contaminants from current or prior land use, disease vectors, or other environmental health conditions that could pose health risks to deployed personnel exist at the deployment sites. Additionally, it also identifies industrial facility operations and commodities near the site that could, if damaged or destroyed, release contaminants that are harmful to personnel. An OEHSA is generally conducted in conjunction with an EBS since the two documents support each other. While the EBS is generally more visual and engineer-related, the OEHSA is more analytical (including a greater variety and detail of sampling), with a greater focus on health hazards.

B-20. ESOH guidance and standards for the operational area are articulated in the CCDR's plans and orders. The CCDR may develop a theater policy for environmental protection and enhancement that is similar to AR 200-1. Base camp planners are responsible for ensuring that these standards are appropriately integrated within mission planning and the base camp development planning process. Initial site selection must include ESOH factors. These include items such as electrical systems; water systems; ventilation; air quality; slip, trip, and fall hazards; structural integrity; and the use of existing industrial infrastructure, such as overhead lifts, chain hoists, and cable systems.

B-21. Certain areas of base camp operation require particular attention to avoid effects on the environment and to protect the health and QOL of the residents. Environmental considerations in the development and operation of these sites include the following:

- Field sanitation.
- HAZMAT storage, transportation, treatment, redistribution or reuse, and safeguarding.
- Spill response and reporting.
- Base camp expansion or contraction potential.
- Petroleum, oils, and lubricants storage, distribution, and safeguarding.
- Integrated waste management, to include the collection, transportation, storage, segregation, recycling, treatment, and disposal of solid waste, black water (sewage), gray water, HW, special waste, medical waste, and explosive waste.
- Maintenance and management of waste management areas and equipment.
- Water conservation, distribution, and reuse.
- Dust abatement.
- Latrine and shower facility locations.
- Dining facility locations.
- Establishment of guidance and policy on ESOH standards.

Appendix B

- Integrated pest management for protection against disease vectors including pesticide use, storage, and disposal.
- Motor pool and maintenance locations.
- Washrack locations and operation.
- Drainage and storm water management.
- Analysis of the threat to base camps

B-22. The staff determines threats to base camps by analyzing threat organizations (manpower, equipment, and available resources), threat patterns, TTP, and support mechanisms. During mission analysis/problem framing and the initial IPB, the staff attempts to—

- Identify emerging threats and TTP.
- Predict threat capabilities and intentions to attack base camps or affect base camp operations, such as the disruption of local contracted support.
- Determine enemy indirect-fire capabilities that can vary at base camps.
- Determine threat abilities to penetrate base camps from underground through tunneling or existing man-made features.
- Determine threat abilities to acquire and use weapons of mass destruction, IEDs, and other weapons that could result in significant casualties or damage on base camps.
- Determine and prioritize information requirements that are fulfilled through information collection activities, requests for information (RFIs), and reachback.
- Identify high-value targets associated with threats, and recommend high-payoff targets for targeting.
- Assess base camp vulnerabilities that can be exploited by adversaries.

B-23. Any uncertainties associated with base camps become information requirements that are fulfilled through information collection efforts, RFIs, and reachback. Information requirements that are of importance to the commander are nominated as commander's critical information requirements. See ADRP 5-0 and MCWP 5-10 for more information.

Determine Specified, Implied, and Essential Tasks

B-24. The staff analyzes the higher headquarters order and the higher commander's guidance to determine specified and implied base camp tasks. The base camp activities are useful in visualizing and organizing the broad range of tasks associated with base camps. From these tasks, combined with the commander's guidance, the staff collectively develops recommended essential tasks for base camps. The results of task analysis lead to the identification of the functional requirements for the base camp, and ultimately its purpose, and are dependent on analysis and input from each staff section or functional area within the command post. These functional requirements and the base camp purpose drive base design.

B-25. Essential tasks are those tasks that the unit must conduct to meet the commander's intent and accomplish the mission. The staff recommends essential base camp tasks during the mission analysis briefing. At the conclusion of the mission analysis briefing, the commander approves those essential tasks considered relevant. A fully developed essential base camp task has a task and purpose.

Determine Available Assets and Shortfalls

B-26. Within the task organization, the staff determines the availability of forces and specialized equipment that can be used to perform base camp tasks. The staff also assesses the amount of construction materials on hand or readily available through the supply system and the ability to perform troop and/or contractor construction based on the availability of manpower, funding, and contracted support. Planners also consider other available support within the operational area or through reachback, to include—

- Joint and multinational engineering, sustainment/logistics, and transportation units.
- Units well-suited for operating and managing base camps, such as Army regional support groups that are augmented with the necessary facility engineering skills and other capabilities.

- Contractors, based on their suitability to work in certain areas given the effects of ethnic, religious, or political boundaries.
- Commercially-available equipment and construction materials that can be acquired through local purchase or contracting.
- USACE, NAVFAC, IMCOM, and the United States Army Medical Command (USAMEDCOM) support and assistance teams.
- Governmental and nongovernmental organizations.

B-27. Based on the specified and implied base camp tasks identified during mission analysis, planners perform a troop-to-task analysis to determine any obvious shortfalls (based on the current task organization) and initiate requests for augmentation through the proper channels. The staff makes assumptions on expected augmentation to facilitate the continuation of planning. It is important to identify any shortfalls or special equipment requirements as early in the planning phase as possible. Long lead times are often required in coordinating for specialized engineer teams, USACE and USAMEDCOM support, and assistance teams since they will likely be in high demand—especially during the initial onset of operations. When considering the use of contracted support, the availability of funds is also a factor. Funding shortfalls must be identified and submitted through appropriate channels as early in the planning phase as possible. When additional support is not available, base camp planners must be prepared to rely on technical expertise available through reachback.

B-28. The availability and accessibility of resources depends on threat conditions, the cooperation of the HN and local government, and the support of local populations, to include their ability to provide skilled labor that is suitable for established standards. Access to in-theater resources, or reserves that can be moved into the theater, effects the planning and design of base camps. In countries or regions with a well-developed infrastructure, materials and skilled labor may be readily available by the relatively easy means of transporting them into the area or through using local resources. If the infrastructure is poor or if the tactical and political situations are unfavorable (as in unassisted and forced-entry situations), resources are more difficult to obtain.

B-29. The relative abundance of certain types of construction materials and the local labor market drive base camp planning and design decisions. The ability to obtain certain construction materials, such as concrete rather than wood products, and the ability of the local labor force to work with those materials may dictate how camps are constructed. Other civilian trades, such as skilled electricians and plumbers, also affect design and construction management decisions.

Determine Constraints

B-30. The staff determines any constraints on base camps. Constraints for base camps may include—
- Allowable design and construction standards in theater-specific guidelines.
- Higher headquarters policies, procedures, OPLANs/OPORDs, and directives.
- Joint and Service directives and regulations.
- International and U.S. laws and regulations, as applicable.
- Construction funding limitations.
- HN laws and local customs and practices.
- Tactical or operational considerations.

Determine Time Available

B-31. The staff determines the time available for planning and ensures that subordinate units are provided approximately two-thirds of the total time available for planning. As planners begin developing possible solutions, they consider the estimated times of arrival for organic and augmenting troops, equipment, and materials that are needed for performing base camp tasks. They also consider the time required for acquiring the necessary funding and approval for local purchases and for contracting services and support.

Appendix B

Determine Information Requirements

B-32. Base camp information requirements are identified collectively, and then selected staff members gather the necessary information within their area of expertise through their respective staff section or through reachback. For example, the engineer staff officer may pursue construction-related information through reachback to USACE support centers, while PVNTMED personnel might coordinate through USAMEDCOM channels. Information is also gathered through information collection to include infrastructure reconnaissance and assessments and through the submission of RFIs to lower, adjacent, and higher units. Some examples of base camp-related IRs include—

- The anticipated life span of a base camp.
- The expected base camp population and events that could cause major fluctuations (surges or reductions), such as transfers of authority and right-seat rides.
- Facility allowances and construction standards.
- The availability of contracted support, funding, materials, and the like.
- Local government and population attitudes on base camps and/or willingness to cooperate and provide assistance.
- Potential constraints, problems, or hazards identified from initial health site assessments and environmental or engineering surveys.

B-33. As with any construction effort, base camp planning is significantly enhanced with on-the-ground reconnaissance and assessments. Whenever possible, predeployment site surveys or on-site reconnaissance is conducted to verify actual conditions and the availability and status of existing facilities. Commanders consider requesting assistance from sources beyond their control, including infrastructure reconnaissance and assessment teams; safety, environmental, and PVNTMED personnel; base camp development teams; and specialized engineer units that are skilled in gathering the information needed for developing base camps.

B-34. Information management is critical to this task. Planners must work together to determine how base camp-related information will be generated, gathered, stored, and disseminated to ensure that the right information is provided to the right people at the right time to facilitate decision making. It is particularly important to avoid sending redundant or irrelevant RFIs to higher headquarters or through reachback and to ensure that RFIs from subordinate units are handled in a timely manner.

Begin Risk Management

B-35. Risk management is the primary process for identifying hazards and controlling risks during operations. Risk management is the process of identifying, assessing, and controlling risks arising from operational factors and making decisions that balance risk costs with mission benefits. For base camps, the staff focuses on health and environmental hazards associated with each aspect of the life cycle. Of particular importance are hazards associated with occupying existing facilities (or areas) that may have residual contamination or be structurally unsound and hazards that are inherent in waste generation and disposal. Other important planning considerations include explosive safety requirements. See ATP 5-19 for a detailed discussion on risk management.

COURSE OF ACTION DEVELOPMENT

B-36. After essential tasks for base camps are approved, the staff integrates them into COA development (and refines them as necessary) while considering the base camp principles. The staff develops associated methods to complete the essential tasks by allocating resources and recommending priorities based on the commander's guidance. The methods are then synchronized to produce the desired effects.

B-37. As described in ADRP 3-0 and ADRP 5-0, lines of operations and lines of effort bridge the broad concept of operations across discreet tactical tasks. Planners may use lines of operations and lines of effort to build their broad concept. If used, planners should consider including base camps within them since base camps represent a fundamental tactic for projecting and sustaining combat power.

B-38. As part of COA development, planners develop a basing strategy for each COA that is linked to the higher headquarters basing strategy. Planners ensure that the necessary assets required for executing essential tasks for base camps are included in the groupings of forces being arrayed, including the requirements for base camp management and operations such as base camp management centers, BCOCs, and BOCs. This allows the COAs to be feasible, from a base camp perspective, while they are being developed.

B-39. As planners array forces based on mission requirements (force ratios and troop-to-task analysis) and commander's intent, they also consider where to best base forces (personnel and equipment) based on the results of the analysis of terrain and civil and environmental aspects considered during mission analysis. Planners must understand the unit capabilities for establishing base camps based on preliminary construction estimates (labor, materials, and time) to ensure that the number of base camps being integrated within a COA is feasible. This ensures that the COAs remain feasible, from a base camp perspective, as they are developed. JCMS is helpful in providing planners with rough estimates on resources and time needed to build standard types of base camps based on purpose and size.

B-40. As COA development progresses, planners make note of any additional base camp requirements and other information that may affect base camp development planning and make the necessary adjustments. For example, the positioning of a base camp close to a populated area may require the construction of a bypass road to avoid interference with local traffic or potentially require T-walls, rather than simply a wire perimeter, due to a lack of depth in the outer security area.

COURSE OF ACTION ANALYSIS/ COURSE OF ACTION WARGAMING

B-41. The staff uses wargaming to test, refine, and adapt the scheme of base camps for each COA. This includes—
- Validating the composition of units (personnel and equipment) performing base camp construction; operations, maintenance, and repair; and base camp management tasks.
- Refining base camp-related tasks to be executed by subordinate units.
- Identifying decision points and triggers for critical, base camp-related tasks such as employing an MSF and increasing levels of capabilities.
- Refining commander's critical information requirements (CCIRs) and IRs and incorporating them into the information collection plan.
- Refining maneuver and fire support plans in support of base camp security and defense missions.
- Refining command and support relationships for base camps.
- Refining the sustainment/logistics plan, including plans for water and energy production and distribution, and waste management, based on the basing strategy or scheme of base camps.

B-42. At a minimum the staff should wargame critical, base camp-related tasks or events, when time for more extensive wargaming is limited. Some examples of critical actions associated with base camps include—
- Timing and sequencing for employing a TCF or another available initial response force to defeat Level III threats.
- Employing first responders in response to a base camp emergency such as—
  - A major fire.
  - An environmental emergency.
  - A mass casualty event.
  - An emergency in a neighboring town or city.
- Conducting a full or partial base camp evacuation.
- Responding to local demonstrations or riots that threaten base camps or affect accessibility.
- Losing a critical ground or air LOC for a base camp.
- Committing traffic control elements to divert and relocate dislocated civilians who are blocking base camp access.

B-43. Once wargaming and COA refinement are complete and a final task organization has been determined, a communications plan can be developed to support each base camp. Establishing an effective communications plan for base camps ensures that they remain responsive to the commander's needs.

ORDERS PRODUCTION, DISSEMINATION, AND TRANSITION/ORDERS DEVELOPMENT

B-44. The staff prepares the order or plan by turning the selected COA into a clear, concise concept of operations and providing the necessary detailed information required by subordinate units for execution. The detailed information that that is lacking from the base order but which subordinate units will need is placed in an attachment (annex, appendix, tabs, and exhibits). Attachments are prepared in a form that best portrays the information, such as text, a matrix, a trace, an overlay, an overprinted map, or a table. See appendix C for guidelines for creating a base camp appendix to an order or plan and FM 6-0 and MCWP 5-10 for more information on orders and attachments to orders.

B-45. Certain base camp-related outputs are generated as a result of mission planning. They include—
- A basing strategy or scheme of base camps.
- Planned base camp locations, the supporting real estate acquisition actions, and the necessary inputs into the information collection plan to drive engineer, infrastructure, and environmental reconnaissance and assessments.
- Tenant and transient unit/organization personnel and equipment lists for each base camp.
- The designated purpose, levels of capabilities, and functional requirements for each base camp to drive facility and infrastructure design and land use planning (site design)—and linkages to other base camps as appropriate.
- Base camp standards that are within theater-established guidelines.
- Designated levels of capabilities that are linked to the basing strategy or scheme of base camps.
- Construction means available (labor, equipment, material, and funding) to facilitate design.
- Base camp construction plans or construction directives.
- Initial life cycle cost estimates and economic analyses.
- Initial IRs and collection plan.
- Initial waste management projections.

PREPARATION, EXECUTION, AND ASSESSMENT

B-46. Base camp planning is continuously refined during preparation and execution as SU/SA improves. Much of the detailed, on-site information needed to finalize base camp design and start construction may not become available until subordinate units have occupied a certain area and/or on-site reconnaissance has occurred. If the situation deviates from the order (for example, a proposed base camp site is unfeasible based on current site conditions), commanders direct adjustments based on staff recommendations and coordinate modifications with their higher headquarters. In some situations, commanders may determine that the current order, to include associated branches and sequels, is no longer relevant to the situation. In these instances, commanders reframe the problem and initiate planning activities to develop a new plan. The planning and design, the preparation, execution, and assessment process of base camps is often a collaborative effort between higher headquarters, constructing units, base camp commanders/BOS-Is, and tenants.

B-47. Throughout the operations process, base camp planners begin planning and coordinating for base camp modifications and improvements to that are needed to mitigate predicted changes in the situation, improve efficiencies, and/or increase the desired level of services. The initiation of construction projects often requires long lead times to acquire the necessary funds, construction materials, project approvals, and contracted support. Base camp planners must often look further in advance than the typical tactical planning windows that are observed.

PREPARATION

B-48. After the order is issued, units begin preparation activities to improve their ability to conduct operations. During preparation, base camp planners collectively monitor ongoing preparatory actions within their respective functional areas, continue gathering and generating base camp-related information, and conduct planning refinement as necessary. Key preparation activities for base camps include—
- Gaining project approval and programming (funding) construction.
- Ordering supplies and materials, with a focus on those with long delivery dates.
- Developing and obtaining requirements packet approval.
- Conducting preconstruction meetings with construction units or contractors.
- Identifying and nominating CORs.
- Refining facility and infrastructure designs based on new requirements or new information resulting from answered IRs.
- Coordinating the linkup of augmenting units/organizations, such as forward engineer support teams and engineer facilities detachments, with the supported unit.
- Planning refinement based upon situation changes and new information. Changes in the situation that could affect base camps include—
  - Revised unit arrival dates, based on deployment and movement timelines, which could affect the availability of base camp assets such as specialized engineer teams or assistance teams.
  - Increases in protection measures based on threats, which could impede movements or restrict contractor access to base camps.
  - Task organization changes that could increase or reduce base camp populations and affect current design capacities.
  - Changes in the availability or status of base camp resources such as existing facilities, contracted support, and shipments of materials.
  - Changes in the concept of operations and the basing of forces into areas not previously considered.
- Supporting subordinate unit base camp planning through collaboration and/or coordination for reachback and responding to RFIs.
- Standing up base camp management centers, BCOCs, and BOCs as prescribed in the OPLAN or OPORD.

EXECUTION

B-49. As base camps are constructed and become operational, commanders and staffs monitor the situation, assess progress, and make adjustments as needed. Commanders continuously assess the progress of base camps based on new information, running estimates/staff estimates, and assessments from subordinate commanders. When the situation deviates from the order, commanders direct adjustments to exploit opportunities and mitigate challenges. Commanders and staffs use the rapid decision-making and synchronization process and the rapid response planning process to make those adjustments and rapidly resynchronize forces and warfighting functions. At any time during the operations process, commanders may choose to reframe the problem and develop a completely new plan when changes in the operational environment render the operational design concept, and the associated understanding and logic behind it, no longer applicable. See FM 6-0 and MCWP 5-10 for more information. During execution, operational commanders and base camp commanders/BOS-Is and their supporting staffs monitor such things as—
- Mission duration changes that affect anticipated base camp life spans and the designated levels of capabilities.
- Task organization and the repositioning of forces changes that affect base camp populations.
- Situations that may prompt base camp realignments, transfers, and closures and can affect the populations of other base camps.
- Threat conditions changes that affect the use of contractors and their access to base camps, their access to local resources, and the overall sustainment and functioning of base camps.

Appendix B

- The status of funding, project approvals, and contracting actions that affect the construction of base camps and operational timelines.
- The status of critical facilities and infrastructure on the base camp.

ASSESSMENT

B-50. Operational and base camp commanders/BOS-Is and their supporting staffs monitor the current situation for unexpected success, failure, or adversary actions that can prevent base camps from progressing toward the desired end state. A large number of base camp life cycle activity management tasks that require significant monitoring, data collection, and assessment must be performed by the base camp commander/BOS-I and staff. Staffs continuously assess the effect of new information on base camp operations. They update their running estimates/staff estimates and determine if adjustments are required. The focus of assessing base camp activities varies during the operations process as follows:

- During planning, assessment centers on developing SU/SA, establishing measures of effectiveness and performance, and evaluating COAs for the commander's decision.
- During preparation, assessment is focused on determining the friendly unit readiness to execute base camp activities and on implementing any refinements to orders based on changes in the threat situation or civil considerations.
- During execution, assessment is aimed at identifying any variances between the current situation and forecasted outcomes. The lessons that units learn while conducting base camp activities are conveyed in TTP.

Note. See FM 6-0 and MCWP 5-10 for more information on running and staff estimates.

B-51. The BOC, base camp working group, master planning working group meetings, and project approval and acquisition review boards play an important role in assessing the overall efficiency and effectiveness of base camps. At the lowest level, self-assessment checklists can be created and distributed to units or individuals, such as facility managers, are assigned specific base camp responsibilities to help assess effectiveness.

Appendix C
Sample Army Base Camp Appendix/Annex

This appendix provides guidelines for creating a base camp appendix/annex as an attachment to an OPLAN or OPORD. See FM 6-0 and MCWP 5-10 for more information on plans and orders, annexes, appendixes, and tabs.

GUIDELINES FOR BASE CAMP APPENDIX/ANNEX

C-1. The base camp appendix/annex contains information, administrative support details, and instructions that expand upon the base order, enabling subordinate unit planning and successful mission execution. The base camp appendix may be part of the engineer or sustainment, depending primarily on the echelon and how the functional responsibility for base camps is organized within the staff.

C-2. The sample base camp appendix shown in figure C-1, page C-2, follows the five-paragraph format for attachments prescribed in FM 6-0 and MCWP 5-10 and should be used as a guideline. The base camp appendix can include any combination of text, matrixes, and graphics to best communicate information to subordinates. Although the content may vary based on unit SOPs and unit skill level, the base camp appendix should meet the following general criteria:

- Contains all critical information and tasks pertaining to base camps not covered elsewhere in the order.
- Does not contain items covered in SOPs unless the mission requires a change to the SOP.
- Provides information that is clear and concise.
- Includes only information and instructions that have been fully coordinated in other parts of the plan or order.

C-3. The base camp staff integrator or base camp working group facilitator ensures that the optimal amount of information and the tools that subordinate units need for base camp planning and execution are provided for in OPLANs and OPORDs. The information needed is generated by various staff members. The base camp staff integrator or base camp working group facilitator is responsible for collecting and consolidating the necessary information into the base camp appendix and ensuring that the information and instructions are consistent with other information contained throughout the plan or order.

Appendix C

[Classification]
(Place the required classification at the top and bottom of every page of the appendix.)

Copy ___ of ___ copies
Issuing headquarters
Place of issue
Date-time group of signature
Message reference number

Include heading if attachment is distributed separately from the base order or higher-level attachment.

APPENDIX ___ (BASE CAMPS) TO ANNEX _____ TO OPERATION PLAN/ORDER NO___.

References: Refer to published theater base camp standards, design guides, and other policies and guidance that apply to base camps.
Time Zone Used Throughout the Order:
1. **SITUATION**. Include information affecting base camp operations that is not covered elsewhere in the plan or order.
   a. **Area of Interest**. Refer to annex B (Intelligence) as necessary.
   b. **Area of Operations**. Refer to appendix 2 (Operations Overlay) to annex C (Operations) as necessary.
      (1) **Terrain**. Describe how the terrain will affect base camps. Include such things as environmentally sensitive areas; areas with historical, cultural, or religious significance; and existing facilities and infrastructure. Refer to tab A (Terrain) to appendix 1 (Intelligence Estimate) to annex B (Intelligence) as necessary.
      (2) **Weather**. Describe how weather will affect base camps. Refer to tab B (Weather) to appendix 1 (Intelligence Estimate) to annex B (Intelligence) as necessary.
   c. **Enemy Forces**. Describe how the enemy will affect base camps. Refer to annex B (Intelligence) as necessary.
   d. **Friendly Forces**. Outline the higher headquarters basing strategy or scheme of base camps. List higher, adjacent, and other functional area assets that support or affect the issuing headquarters base camp capabilities or require coordination and additional support. Refer to annex P (Host-Nation Support) as necessary.
   e. **Interagency, Intergovernmental, and Nongovernmental Organizations**. Identify and describe other organizations in the area of operations that may affect base camps. Refer to annex V (Interagency Coordination) as necessary.
   f. **Civil Considerations**. Describe the effects of civil considerations on base camps. Refer to annex K (Civil Affairs Operations) as necessary.
   g. **Attachments and Detachments**. List units attached or detached only as necessary to clarify task organization. Refer to annex A (Task Organization) as necessary.
   h. **Assumptions**. List any base camp-specific assumptions that support the appendix development.
2. **MISSION**. State the mission of base camps or base clusters in support of the base plan or order.
3. **EXECUTION**.
a. **Scheme of Base Camps**. Describe how the commander intends to use base camps to support the concept of operations. Describe the overall arrangement of base camps/clusters, and clarify the interrelationships (hub and spoke) as necessary. State the priorities for base camps (by unit or area) for each phase of the operation (if the operation is phased), to include base camp transfers and closures. Supplement the concept of sustainment (paragraph 4 of the base order) with any additional information that clarifies base camp tasks and purposes.

[page number]
[Classification]

Figure C-1. Sample Army base camp appendix/annex

b. **Tasks to Subordinate Units.** List base camp tasks that are assigned to specific subordinate units but not contained elsewhere in the plan or order.
c. **Coordinating Instructions.** List instructions that apply to two or more subordinate units but not covered elsewhere in the plan or order. This may include, but not be limited to—
- Base camp standards.
- Construction programming and funding procedures.
- Project approval and acquisition review procedures (with thresholds specified as necessary).
- Transfer and closure procedures.
- Environmental, safety, and occupational health measures for reducing risks associated with constructing and operating base camps.
- Disposition or disposal instructions for construction debris and discarded materials.
- Base camp-related information requirements. Including requests for information that may be relevant to subordinate unit planning and have been submitted to higher and adjacent units.
- Channels for contacting support (reachback) for technical assistance.
- Instructions for disseminating base camp-related information.
- Master planning requirements.

4. **SUSTAINMENT.** Identify priorities and their key tasks for sustainment for base camps, and specify additional instructions as required. Describe stock levels or basic loads for construction and barrier materials and other base camp-related items to be maintained at each base camp. Describe the appropriate channels for ordering, acquiring (including local purchases), and contracting base camp supplies, materials, and services that are not covered in annex F (Sustainment) as necessary. Clarify any support requirements (for transient units or daily base camp visitors) or means for receiving support on an area basis that are not clearly articulated in annex F (Sustainment).

5. **COMMAND, CONTROL AND SIGNAL.**
   a) **Command.** Identify base camp/cluster commanders, and clarify command and support relationships to ensure unity of effort for base camps. Clearly identify approving authorities (and thresholds, as applicable) for base camp construction programming and funding, project approvals, and acquisition reviews. State the location of key personnel involved with base camps.
   a) b. **Control.** Describe the employment and location of base camp management centers, base cluster operations centers, and other centers as known. State any base camp liaison requirements not covered in the base order.
   b) c. **Signal.** Address communications requirements and reports used for operating and managing base camps. Refer to annex H (Signal) as appropriate.

ACKNOWLEDGE: Include only if distributed separately from the base order.

OFFICIAL:
[Authenticator's name]
[Authenticator's position]
Either the commander or the coordinating staff officer responsible for base camps may sign the appendix.

TABS: List any tabs as required. Tabs may include—
- Base camp standards.
- Base camp transfer and closure procedures.
- Base security and defense.
- Master planning procedures.
- Base camp construction plans and construction directives.
   DISTRIBUTION: Show only if distributed separately from the base order or higher-level attachment.
[page number]
[Classification]

Figure C-1. Sample of Army base camp appendix/annex (continued)

This page intentionally left blank.

Appendix D
Base Camp Land Use Planning

Land use planning is the science of assigning suitable usage to parcels of land. The preparation of the plan takes into account many factors including requirements, interrelationships, constraints, and future expansion. The final, approved product should plan for comparable land uses to be arranged, close to each other while are incompatible land uses are not.

LAND USE PLANNING

D-1. Land use planning, also commonly referred to as site design or base camp layout, is the process of calculating, mapping, and planning the allocation of land areas. This planning is based on—
- Land use categories.
- Terrain characteristics (lay of the land).
- Operational requirements.
- Functional interdependencies (affinity relationships).
- Protection, civil, and environmental considerations (standoff and separation between facilities).
- Base camp standards and commander's guidance.

D-2. Land use planning is an art and a science in logically arranging the required facilities and infrastructure within a specific site with the least negative effect to the environment and the greatest benefit to users. The land use plan provides the framework for the layout of the base camp and is updated through the master planning process. The steps of land use planning are detailed in EP 1105-3-1. The general categories are—
- Collect information.
- Set land use goals and objectives.
- Calculate land area requirements.
- Conduct an environmental analysis.
- Prepare an environmental overlay.
- Conduct a functional analysis.
- Produce a functional relationship overlay.
- Develop alternative land use plans.
- Select the best alternative land use plan.
- Obtain the commander's approval.
- Implement and maintain the land use plan.

D-3. In developing the land use plan, planners consider—
- Tenant and transient unit facility requirements, to include adequate space for unit system operations and maintenance, storage, training, and expansion.
- Affinity relationships (functional interrelationships between facilities). See ATP 3-37.2 for more information.
- AT/force protection measures include—
  - A layered security approach that is focused on the protection of critical assets with and has adequate dispersion and standoff.
  - Standoff distances or geographic isolation to minimize the accessibility and vulnerability of critical facilities such as ammunition/explosives, POL, and HAZMAT/HW storage areas.

Appendix D

Note. See GTA 90-01-011, TM 5-304, and TM 38-410 for standoff distances and separation for structures.

- Explosive safety quantity-distance requirements.

Note. Certificates of risk acceptance may need to be developed to document risk acceptance of violations of required quantity-distance/safe zones. See DA Pamphlet 385-30 and DA Pamphlet 385-64 for detailed instructions.

- Terrain and weather effects (elevation, slope, surface drainage, trafficability, prevalent wind direction).
- Environmental considerations such as standoff from environmentally sensitive areas and separation between food storage or water sources and waste management areas. Depending on the amount of land area available and consideration of the mission variables, some waste management areas, such as sewage lagoons and trash burial and burn sites, may need to be established outside of the perimeter with the necessary protection measures to mitigate any risks.
- Separate working and living areas as required for HN military and government personnel and non-CAAFs that can affect overall utility requirements.
- Accessibility and road requirements, including building spacing to facilitate maintenance, repair, and the movement of first responders.
- Base camp functions and purpose (allowances for expansion and surges).
- Considerations for expansion and partial transfers or closures.
- Utility corridors to facilitate the expansion and repair of utility systems.
- Waste management requirements.

D-4. Although land use planning begins in the early stages of base camp development, it requires planners to conduct a facility requirements analysis before it can be finalized. Additionally, since land use is directly linked to the base camp location selected during planning, planners should confirm that the location is approved and remains suitable based on planning refinement and changes in the situation to include the results of real estate acquisitions. Land use planning can be enhanced through geospatially-referenced software applications, such as GeoBEST, that can interface with JCMS. An example of a land use plan is shown in figure D-1. See appendix F in EP 1105-3-1 for other examples of land use plans.

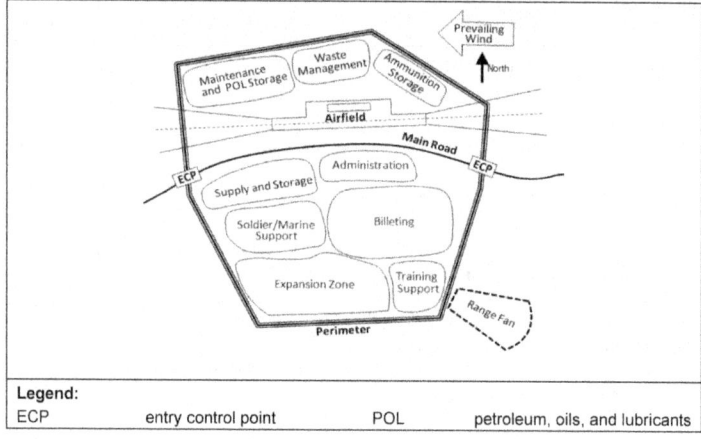

Figure D-1. Sample land use plan

BASIC DESIGNS

D-5. The AFCS provides two standard designs that can be adapted to a particular site—the rectangular box design (see figure D-2) and the wheel design (see figure D-3, page D-4). Although the rectangular box design is more traditional, the wheel design may offer better space utilization, security, and capacity for expansion. Planners must compare the advantages and disadvantages that each design offers based on consideration of the size of the base camp, space limitations, and threat and vulnerability assessments. See TM 5-304 for more information on the AFCS.

D-6. Battalion/battalion landing team and BCT/RCT-size base camps and associated constructing units should have the software and hardware to access standard design databases to produce site-adapted designs that are geospatially referenced, create plans and specifications for construction or contracting, and maintain master plans and as-built drawings. These larger base camps will also typically provide support to platoon and company base camps with which they have a hub-and-spoke relationship and maintain those records as well.

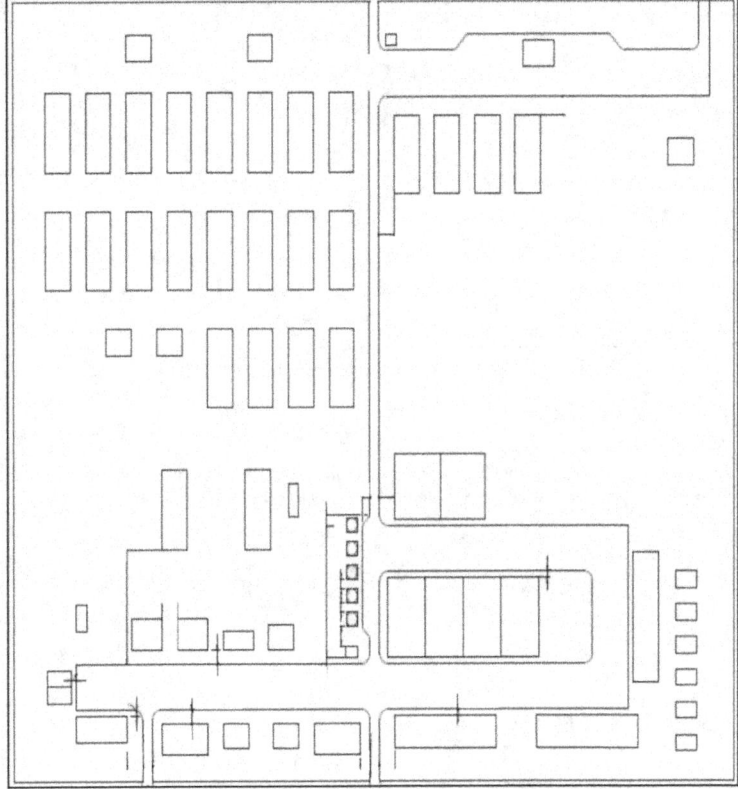

Figure D-2. Rectangular box design

Appendix D

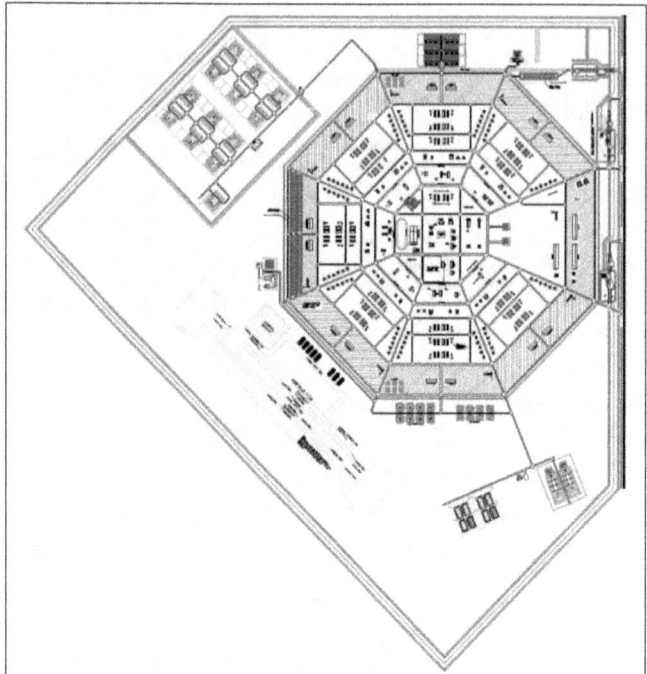

Figure D-3. Wheel design

LAND USE CATEGORIES

D-7. Establishing zoning within a base camp helps to maintain adequate standoff distances, manage traffic and prevent vehicle accidents (with pedestrians and other vehicles), and maintain the sanctity of troop billeting areas. Typical land use categories used for base camps are shown in table D-1.

Table D-1. Typical land use categories for base camps

Land Use Category	Types of Facilities
Operational	- Airfields. - Unmanned aircraft system landing strips. - Landing zones.
Training support	- Training areas. - Weapons-firing ranges.
Billeting	- Tent pads and barracks buildings.
Administration	- Unit headquarters. - Administration buildings. - Communication facilities.
Soldier/Marine support	- Medical treatment facilities (medical, dental, and veterinary services). - Dining facilities. - Laundries. - Barbershops. - Post exchanges and food courts. - Morale, welfare, and recreation facilities. - Fitness facilities. - Chapels. - Education centers.
Nonhazardous material storage	- Warehouses. - Space for the placement of military vans or containers.
Hazardous material, hazardous waste	- Petroleum, oil, and lubricant storage areas. - Ammunition and explosives storage areas. - Hazardous waste accumulation points.
Motor pool/vehicle parking	- Vehicle maintenance facilities. - Specific parking areas for nontactical vehicles.
Utilities	- Facilities for power, water, and waste treatment/disposal. - Rights of way or easements.
Security and defense	- Guard towers. - Entry control points. - Staging areas for response forces with entry and exit points.

ENVIRONMENTAL CONSIDERATIONS

D-8. Environmental considerations begin during planning as part of site selection and continue throughout the base camp life cycle. During the design phase, base camp planners analyze existing EBS (if available) to better understand the initial site conditions and facilitate base camp layout. An on-site investigation is conducted whenever possible to help assess environmental and safety hazards and confirm the overall suitability of the area. Factors such as evidence of environmental contamination, landfills or trash burial sites, and surrounding land uses (industrial complexes) are considered. The existing facilities and infrastructure and the surrounding area are surveyed to help planners determine the best locations for troop billeting, maintenance, HAZMAT and petroleum, oil and lubricant storage, waste management areas, and motor pool locations from an environmental and health perspective.

Appendix D

D-9. While all base camp layouts are unique due to variables such as terrain, threat situation, base camp purpose, and the number and the type of tenant units, certain relationships between base camp layout and environmental considerations tend to be constant. Environmental considerations with regard to base camp layout include—

- Locating petroleum, oil and lubricant and HAZMAT storage areas, HW accumulation points, and motor pools away from billeting areas and drainage features.
- Locating latrines, wastewater treatment sites, trash burial sites, incineration sites, and other waste management areas downwind and away from dining facilities, food storage areas, populated areas, and sources of water.
- Avoiding naturally and culturally sensitive sites.
- Avoiding low-lying areas that are prone to flooding or areas with standing water.

Appendix E
Base Camp Planning Factors

This appendix discusses base camp planning factors, are rules of thumb. Planners use these to estimate requirements and capabilities as a part of running estimates/staff estimates and to help develop basing strategies or schemes of base camps. Base camp considerations or other factors that should be incorporated in planning are discussed in chapter 2.

USE OF PLANNING FACTORS

E-1. Planning factors provide a starting point for preparing running estimates/staff estimates when actual situational data is unclear or unknown, or in the absence of specific policy, planning, or guidance. A rule of thumb is a means of estimation, a general guideline, or a procedure that is easily applied based on experience and common knowledge for approximately calculating—that may not apply to every situation.

E-2. Planning factors and rules of thumb are not intended to be precise or prescriptive. Adjustments must be made based on the uniqueness of each situation. Making adjustments increases the chances of preparing good basing strategies, schemes of base camps, master plans, and base camp development site plans based on estimates of requirements and capabilities.

E-3. Planning factors may be provided in base camp policy, standards, and planning guidance. For example, you could plan to provide one fitness center of 1,024 square feet per 150 authorized users at enhanced company size base camps or one fitness facility at 3 square feet per authorized user at expanded company size base camps. These two different planning factors, applied to the base camp populations associated with each size of base camp (see table E-1, page E-2), would yield estimated facility sizes ranging from 900 square feet to 13,647 square feet (depending on the value of the population range used). The staff uses planning factors to help estimate life cycle requirements, develop COAs, and generate potential solutions to problems. Life cycle requirements are based on changes and adjustments to populations, levels of services and support (quantities and types), size, facilities and infrastructure, real property (land and permanent improvements to the land, including all nonrelocatable buildings), construction estimates of materials, labor, time, equipment, staffing, and cost.

E-4. The CCDR's base camp standards for the AOR may be used as planning factors to complete initial running estimates/staff estimates unless specific base camp guidance has been provided based on the contingency basing strategy for the operational area.

E-5. Base camp planning factors are used for conceptual and detailed planning. Because planning is continuous, the planning factors used to develop conceptual and initial detailed plans are assessed, validated, or updated as the operation progresses, data becomes available, and situational understanding improves. Base camp planning documents are continuously revised, updated, and distributed accordingly.

E-6. Planning factors are recorded and updated, as appropriate, in running estimates/staff estimates. See FM 6-0, JP 5-0, and MCWP 5-10 for more information.

BASING STRATEGY PLANNING FACTORS

E-7. If authorized, any echelon commander may decide to conduct operations from base camps or direct the development of plans for the possible future use of base camps. In the absence of actual numbers, planning factors are used to develop estimates of information needed for the basing strategy or scheme of base camps. Based on the proposed task organization and estimated personnel strengths, the engineer and sustainment planner would normally work with the operations officer to develop an initial estimate of the required number, size, duration, and possible commanders of base camps across the AO. The engineer and intelligence

Appendix E

planners would begin assessment of possible site locations within proposed unit boundaries. The sustainment/logistics planner would work with the operations officer and engineer to develop an initial array of sustainment/logistics hubs, base camps with major support and service missions, and the initial level of services for the base camp. Possible base camp hub-and-spoke options for protection, security and defense, sustainment/logistics, or communications support could be analyzed.

E-8. Base camp commanders/BOS-Is and staff for most base camps can be dual-hatted or based around a unit assigned a nontraditional mission and assigned/task organized to a base camp for that purpose.

E-9. For example, using the base camp sizes and population planning factors shown in table E-1, a five-brigade or regimental-size force with an estimated strength of 15,000 might operate from one or several base camps. Several COAs should be developed and analyzed. A regional support group or MEB (Army) can provide the camp commander/BOS-I and staff to the support area base or function as a base cluster commander located at one of the base camps. Each would require staff augmentation based on the specifics of the base camp or network of base camps that they are responsible for. Most other base camp commanders/BOS-Is would be dual-hatted, perhaps with one or more commanded by a unit performing a nontraditional mission. Estimates of joint, interagency, intergovernmental, and multinational personnel and daily visitors are arrayed across the matrix of camps to get a more accurate estimate of total camp populations, with an attempt being made to reflect populations over time, or phases if the operation is phased. The final plan would be obtained through an iterative process of analysis and tradeoffs.

Table E-1. Base camp sizes and planning factors

Base Camp Size	Approximate Population	Dimension	Surface Area Required (not including standoff)	Length of Perimeter (nominal)
Platoon	50	150 meters by 250 meters	37,500 square meters	800 meters
Company	300	300 meters by 450 meters	135,000 square meters	1,500 meters
Battalion/battalion landing team	1,000	500 meters by 1,200 meters	600,000 square meters	3,400 meters
Brigade/regimental combat team	3,000	To be determined by base camp planners	To be determined by base camp planners	To be determined by base camp planners
Support area	6,000 or greater	To be determined by base camp planners	To be determined by base camp planners	To be determined by base camp planners

E-10. The initial plan should assume that all base camps are continuously operated, with a commander and staff and provided with required sustainment/logistics and protection functions. As the security situation improves, some base camps could be mothballed or operated with a skeleton staff and opened and occupied as required–or ultimately prepared for closure.

E-11. Planners use base camp planning factors to estimate the total sustainment/logistics requirements; facility requirements; balance between troop and contracted construction, support, and services; total supply space and shipping cube space requirements or local procurement strategy: contract methods, capacities, quality control, and material delivery schedule.

E-12. One base camp should be designated as a regional command and/or sustainment/logistics hub that has a C-130-capable or larger airfield. Linkages from this base camp to major LOCs should be included as part of the master plan.

E-13. Planning factors for specific types of facilities, designs, construction, protection, security and defense, and sustainment/logistics estimating are found in the related doctrinal manuals. ADRP 4-0 discusses the sustainment estimate and use of the Web-based Operations Logistics Planner to assist in developing the project. Although not base camp-specific, the tool uses the latest Army-approved planning rates. A similar Web-based tool could be developed for base camp engineering estimates. The UFC system provides

planning, design, construction, operations, and maintenance criteria and applies to all service commands having MILCON responsibilities. UFC are effective upon issuance and are distributed only in electronic media from the following sources:
- UFC index.
- USACE technical information.
- The Construction Criteria Base System maintained by the National Institute of Building Sciences.

LAND USE PLANNING FACTORS

E-14. The base camp development planning process may be adjusted based on specifics associated with climate, and temperatures, associated with the given AO. Land use planning factors include the following:
- Land area requirement calculations.
- Facility standoff/separation.
- Location selection.
- Facility requirements.
- General site planning.
- Design guides and programming.
- Maintenance.
- Painting.
- Roads.
- Ground cover.
- Preventive maintenance.
- Cleanup, Closure, and archive.

Land Area Requirement Calculations

E-15. Table E-1 provides base camp dimensions, the surface area required, and the length of perimeter for each base camp size and population range. The dimensions are approximate based on AFCS initial standard designs. The surface area required and length of perimeter are based on the dimensions. The dimensions in this table are minimums used for initial planning and do not include all facility standoff/separation for health, safety, AT/force protection, or expansion requirements discussed below. These dimensions must be adjusted based on other considerations such as the threat, usable terrain, waste management areas, and ammunition/explosive storage.

E-16. An estimate of area requirements is primarily based on base camp population, security area standoff requirements, and an expansion zone. The surface area planning factors shown in table E-1 should be adjusted for security standoff requirements of each base camp. Base camps in a higher-threat area may require greater facility dispersion and protective measures. The surface area of the base camp must be adequate for accommodating anticipated expansion throughout the life cycle. The area required, based on population, is adjusted after completing estimates of facility and protection requirements (for example, depth of the outer security area). Some unique facilities may require extensive land, such as an unmanned aircraft system runway. See engineer doctrine for land-planning factors for waste management areas.

Facility Standoff/Separation

E-17. There are several planning factors for standoff/separation of facilities. See UFC 4-010-01, UFC 4-010-02, ATP 4-25.12, and ATP 3-34.5/MCRP 3-40B.2 for standoff planning factors for waste management areas.

Location Selection

E-18. Location selection is a balance of operational, sustainment, and construction requirements. Some rules of thumb for site location selection include the following:
- All sites considered as potential base camps sites should be scalable and easily expanded.
- The most desirable site locations are those that are easiest to secure and defend.

Appendix E

- Whenever the establishment of a base camp is being considered, at least three suitable, possible locations (COAs or options) should be identified before recommending the most advantageous COA.
- The entire staff should be involved in evaluating potential base camp sites.

Facility Requirements

E-19. Facility requirements integrate facility allowances with supported and/or tenant unit requirements. Following are some rules of facility requirements:

- A prioritized list of projects for initial construction, O&M, and follow-on improvements should be developed in the master plan to mitigate the fact that requirements always exceed resources.
- All requirements that exceed standards should be approved at least at the next higher headquarters.
- Priority for fulfilling facility requirements should be: U.S.-owned-occupied, or-leased facilities; HN government support; leased facilities; pre-positioned facilities in-theater; contract construction; and troop construction.

General Site Planning

E-20. General site planning takes the initial land use plan, facility requirements, and unit requirements into account. Following are some rules of thumb for general site planning:

- Be aware that poor site layout can degrade physical health, reduce coordination and cooperation among units, erode morale, and increase operational costs.
- First, lay out land use categories, facility groups, and major interior roads/links to LOCs, then primary utility distribution lines, and finally individual facilities.

Design Guides and Programming

E-21. Design guides and programming ensure that base camps will be functional at the appointed time. The following are some rules of thumb for design guides and programming:

- Design guides should be provided for all base camps.
- MILCON programming timelines do not normally support initial contingency construction requirements.

Maintenance

E-22. The maintenance planning factors discussed here are part of the base camp O&M—not unit equipment maintenance. Base camp maintenance requirements are integrated into the master plan. Maintenance programs and projects should be developed, planned, prioritized, programmed, and monitored. Service components plan for and program funding for maintenance. Tenant units or organizations are normally responsible for the installation and maintenance of all unit-specific items, such as a signal/communications unit satellite dish or an Army and Air Force Exchange System cooler or stove. The goal of routine maintenance is to maximize the life expectancy of facilities and infrastructure with minimal cost. Invest in capital maintenance, repair, and minor construction only for minimum-essential, high-priority, and self-amortizing requirements. Planners should assess HN and contractor maintenance support capabilities. Potential design solutions should be evaluated to reduce maintenance costs.

Painting

E-23. Wooden buildings should be primed and painted to prevent weather damage. Paint should have a durability rating of at least 5 years. Plan to paint interior walls every 18 months.

Roads

E-24. The goal is to maximize maneuverability, minimize damage to equipment, and provide a safe transportation system. Plan for a 30-foot roadway width with 15-foot clear space on either side of the road for utility distribution lines, drainage features, and pedestrian flow. Routine grading is required to maintain drainage and to prevent potholes and washboarding. Minimize the use of loose rock greater than 1.5 inches in diameter on roads and parking lots. Plan routine dust abatement and mud, snow, and ice removal based on local conditions. Paving should be considered on gravel roads if the payback period is 2 years or less.

Ground Cover

E-25. Planting of ground cover, such as native grasses or low-growing plants, reduces dust and erosion. Ground cover that requires minimal watering and mowing should be selected. Mowing should be done under the guidance of the environmental officer based on local conditions for vector control and ground cover survival.

Preventive Maintenance

E-26. Preventive maintenance emphasizes the cyclic and seasonal inspections of systems and facilities. Preventative measures are implemented to preserve the system rather than replace a failed system. The goal is to identify safety issues and reduce the cost by identifying deficiencies while they are still small and easy to fix. Preventative maintenance is also a means of encouraging personnel to be constantly aware of energy conservation measures to implement them.

Cleanup, Closure, and Archive

E-27. Maintenance for base camps scheduled for closure should be limited to emergency or breakdown repairs. Base camps scheduled for transfer should be maintained to the agreed-upon turn over standards.

E-28. Cleanup and closure plans and procedures for made at the start of an operation and integrated into the planning process will help avoid or reduce future challenges. Without other guidance, land should be returned to its original use. Estimate the time to clean up and close a base camp as 50 percent greater than the time to construct it.

FACILITY AND INFRASTRUCTURE DESIGN FACTORS

E-29. One of the possible options for base camp designs using new facilities should be the standard AFCS designs. As plans are finalized, the standard designs are site-adapted. If some or all existing facilities are used, the information from the AFCS can be used as planning factors to help estimate and assess facility requirements.

E-30. Table E-2, page E-6, is a table of sample contingency standards that can be used as recommended minimum planning factors to estimate the type and total requirements of facilities within the theater. These samples do not include all facilities.

E-31. Table E-3, page E-7, is a table of sample contingency design requirements that can be used as planning factors to estimate the type, size, and total requirements of the listed facilities. All area measurements are annotated in net square feet. Net square feet is defined as the usable area available for use by the individual or activity.

E-32. For beddown facilities, a sample planning factor of the recommended minimum area (in square feet) for personnel accommodations using a temporary construction standard is shown in table E-3. These planning factors could later be established as theater standards. The table also shows how many personnel are housed in a SEA hut/SWA hut or container.

PLANNING FACTORS OF FORCE PROVIDER CAPABILITIES

E-33. Each force provider expeditionary module supports 150 personnel with complete self-contained, climate-controlled billeting; quality food and dining facilities; hygiene systems; and MWR facilities. 150-person modules can be linked together to support base camps of any larger size. Components include—
- Air beam-supported tents.
- Energy-efficient insulating liners.
- Solar shade systems.
- Micro grids for more efficient power generation/distribution.

Appendix E

- Shower and laundry water reuse and conservation.
- Light-emitting diode lighting.
- Expeditionary hygiene systems.

Table E-2. Sample contingency standards

Facility	Contingency			Recommended Design Size
	Initial		Temporary	80–300 NSF per person
	Organic <90 Days	Initial <6 months	Organic <90 Days	
Housing	Unit tents[1]	Unit/FP tents	Unit/FP tent to SEA hut/SWA hut	One fixture per 20 personnel
Latrine	Burn-out	Chemical	AB units/SEA hut/SWA hut	One shower head per 20 personnel
Shower	Shower unit tent	Shower unit tent	AB units/SEA hut/SWA hut	Lagoon: 1 acre per 200 personnel
Sewage disposal	Leach field/lagoon	Leach field/lagoon	Lagoon or treatment plant	60–300 NSF per person
Office	Unit tents[1]	Unit/FP tents	SEA hut/SWA hut or container	700 NSF per 1,000 personnel
Helipad	Stabilized earth	Airfield matting	Concrete	Not applicable
Fuel	Bladder[2]	Bladder[2]	Bladder[2]	Not applicable; include secondary containment
Vehicle hard stands	Stabilized earth	Gravel	Concrete	As required
Storage	Unit tents[1]	MILVANs	MILVANs	As required
Roads and streets	Stabilized earth	Gravel[3]	Gravel[3]	Not applicable
Potable water	Bottle	Bottle/WPS	Well, treatment plants	Not applicable
Nonpotable water	Local source	Local source	Local source	Not applicable
Washrack	None	Gravel	Gravel[3]	To accommodate the largest vehicle
Electric	Unit tactical power system[4]	Prime power system[4]	Sustained power system[4]	Not applicable
Dining facility	Unit tent[1]	Unit/FP tents	Unit/FP tents to SEA hut/SWA hut	1,290 NSF per 100 personnel
PX warehouse	Unit tent[1]	Unit/FP tents	Metal prefabricated	4,480 NSF per 1,000 personnel

Notes. Improvements to facilities are dependent on operational situations.
[1] Unit tents are provided by the Service components.
[2] Requires secondary containment.
[3] Requires oil-water separator.
[4] See appendix F for power system descriptions

Legend:
AB	absolution	PX	post exchange
FP	force provider	SEA	Southeast Asia
MILVAN	military van	SWA	Southwest Asia
NSF	Net square footage	WPS	water purification system

Table E-3. Sample planning factors for personnel accommodations for temporary standards

Category			NSF	Number Per SEA/SWA Hut	Number Per 8x20 feet Container
Army	Marine	Civilian			
Private, private first class, specialist, corporal, and sergeant	Private, private first class, lance corporal, and corporal sergeant	GS-05 and below	80	6	2
Staff sergeant, sergeant first class, warrant officer one, chief warrant officer two, first lieutenant, and second lieutenant	Staff sergeant, gunnery sergeant, warrant officer, chief warrant officer two, first lieutenant, and second lieutenant	GS-06 through GS-09	90	5	2
First sergeant, master sergeant, chief warrant officer three, and captain	First sergeant, master sergeant, chief warrant officer three, and captain	GS-10 and GS-11	90	4	2
Chief warrant officer four and major	Chief warrant officer four and major	GS-12	100	4	2
Chief warrant officer five, command sergeant major, and sergeant major	Chief warrant officer five, lieutenant colonel, sergeant major, and master gunnery sergeant	GS-13 and GS-14	125	2	1
Colonel	Colonel	GS-15	150	2	1
Brigadier general	Brigadier general	Senior executive service	300	1	1
Legend:					
GS general schedule		SEA Southeast Asia			
NSF net square feet		SWA Southwest Asia			

DOG KENNELS

E-34. Kennel planning factors include 145 square feet per dog for interior facilities, including kitchen, tack room, and interior dog run (36 net square feet per dog) and 48 net square feet per dog for exterior dog runs. See ATP 3-39.34 for more information.

BUNKERS

E-35. The planning design factor for bunkers and fighting positions is 110 percent of camp population. For the normal planning factor, 50 percent of the population is expected to be on the perimeter during an attack, with 50 percent in bunkers. For more information on bunker and protective structure design, visit the USACE Protective Design Center Web site.

CONSTRUCTION FACTORS

E-36. Construction planning factors include unit, equipment, and personnel capabilities. These factors assist in determining construction support for each COA. LOGCAP program planners can provide planning factors and estimates of contractor capabilities for construction and services. The proximity of a suitable

Appendix E

base course material source is a critical planning consideration. Key metrics for alternative road and airfield plan COA comparison are the total amount of earthwork and number of drainage facilities. Planners assess the availability and capabilities of construction units and contractors to complete the required base camps according to the initial planning schedule and then estimate the operations and maintenance capacities. Professional engineers should approve nonstandard designs and preferably manage base camp construction. See ATP 3-34.40/MCTP 3-40D for engineer unit construction capabilities.

E-37. Tier levels for tent improvements are as follows:
- Tier I consists of a general-purpose, medium field tent or equivalent tent, extendible, modular, personnel TEMPER (16 feet by 32 feet) with plywood floor panels.
- Tier II consists of a general-purpose, medium field tent or equivalent TEMPER with full wooden frame, two electric light outlets, two electrical outlets, wooden sidewalls, and space heaters.
- Tier III consists of a general-purpose-medium field tent or equivalent TEMPER, with plywood panel sidewalls, raised insulated flooring, four electric light outlets, eight electrical outlets, and ECUs.

E-38. Table E-4 is a table of how staff planners can use unit personnel estimates to determine the maximum number of structures required by unit. It also allows planners the ability to compare the total estimated maximum cost.

Table E-4. Sample maximum numbers of personnel structures

	Personnel Required	Personnel on Hand	Maximum Number of Structures Required	SEA/SWA hut Cost	Tent Cost
BCT 1	3,452	3,395	340	$1,773,433.20	$615,566.60
BCT 2	4,207	4,213	421	$2,195,927.58	$762,216.29
BCT 3	4,085	4,386	439	$2,289,815.22	$794,805.11
BCT 4	1,625	1,577	158	$824,124.84	$286,057.42
Sustainment brigade	920	822	82	$427,710.36	$148,460.18
Field artillery brigade	970	978	98	$511,166.04	$177,428.02
Reconnaissance squadron	1,308	1,278	128	$667,645.44	$231,742.72
Engineer battalion	2,463	2,301	230	$1,199,675.40	$416,412.70
Light infantry battalion	2,351	2,351	235	1,225,755.30	$425,465.15
Other division troops	3,304	3,704	370	$1,929,912.60	$669,881.30
Totals	24,685	25,005	2,501	$13,045,165.98	$4,528,035.49
SEA hut/SWA hut cost:	$5215.98		Tent cost:	$1810.49	
Legend: BCT brigade combat team SEA Southeast Asia SWA Southwest Asia					

Transient, Surge, and Contractor Housing

E-39. Base camp designs include planning factors to accommodate for transient, surge, and contractor housing. The expansion of camp footprint should be triggered when the population exceeds original burdened population design factors by the following percentages:
- Platoon – 25 percent.
- Company – 20 percent.
- Battalion – 15 percent.

E-40. The demand and consumption of supplies on hand increases as the population increases. See table E-5 for expansion increase planning factors.

Table E-5. Expansion increase planning factors

Characterization of Occupant	Characteristic	Typical Duration	Base Camp Expansion Planning Factor
Transient	Individuals who are waiting for transition or performing temporary duties and are billeted on the base for a short duration.	14 days or less	Platoon: 10%, Company: 15% Battalion: 20% BCT/RCT: 35% Support area: <35%
Surge	Announced (can be planned for) small groups or units assigned to augment the base or tactical mission for extended periods.	2–52 weeks	Platoon: 25%, Company: 20% Battalion: 15% BCT/RCT: 15% Support area: 10%
Contractor	Nongovernmental civilian personnel employed to conduct specific tasks on or in the general vicinity of the base camp.		Platoon: 10%, Company: 10% Battalion: 30% BCT/RCT: 35% Support area: 50%

Legend:
BCT   brigade combat team   RCT   regimental combat team   %   percent

Unit Headquarters

E-41. Table E-6 contains sample planning factors for unit headquarters maximum space at a battalion size camp. It can be used to assess existing facilities.

Table E-6. Planning factors for unit headquarters at a BCT/RCT size base camp

Unit	Maximum NSF/Unit Headquarters
Brigade	5,376
Battalion	3,840
Company	1,536

Legend:
NSF   net square feet

Unit Facilities

E-42. Table E-7, page E-10, is an example planning factors for unit private and open office space. Table E-8, page E-10, and table E-9, page E-11, are examples of comparison of the total base camp housing cost with the maximum structures required compared to the total cost for GP medium tents and SEA huts/SWA huts.

Medical Treatment Facilities

E-43. Table E-10, page E-11, is an example for basic planning of base camp medical treatment facilities requirements. The actual requirement is directly related to the medical and dental mission and care expectations of the operational command that should be coordinated with supporting medical headquarters staff. Planning should accommodate the medical unit present on the camp whether it is a troop medical clinic, battalion aid station, or a brigade medical support company and the medical support role it provides as defined by FM 4-02 the medical company TTPs and/or ATP 3-37.2.

Appendix E

Table E-7. Example of planning factors for office space

Type of Office	Personnel			Maximum NSF/Person
	Army	Marine	Civilian	
Private	Brigadier and major generals	Brigadier and major generals	SES	300
	Colonel, lieutenant colonel, task force command sergeant major, and chief warrant officer five	Colonel, lieutenant colonel, and task force command sergeant major	GS-15	200
	Lieutenant colonel, major, and brigade/battalion command sergeant major	Lieutenant colonel, major, and brigade/battalion sergeant major	GS-13 and GS-14	150
	Major, captain, sergeant major, and first sergeant, chief warrant officer three–four	Major, captain, sergeant major, and first sergeant	GS-12	100
Open	Captain, first lieutenant, second lieutenant, warrant officer 1–2, first sergeant, and master sergeant	Captain, first lieutenant, second lieutenant, warrant officer one, first sergeant, and master sergeant	GS-09 through GS-11	110
	Sergeant first class	Gunnery sergeant	GS-07	90
	Administrative and clerical positions	Stenographic and clerical positions	Stenographic and clerical positions	60

Notes.
1. Applies *only* to military units or organizations and personnel. Administrative space for MWR and commercial functions are discussed separately.
2. To calculate the total building size, add an additional 40 percent for central files; hallways; and storage, copiers, mail, and conference rooms.

Legend:	
GS	general schedule
MWR	morale, welfare, and recreation
NSF	net square feet
SES	senior executive services

Table E-8. Maximum average estimated cost for general purpose medium base camp

Item	Unit Cost	Quantity	Total Cost
Housing (tent, general purpose, medium)	$1,810.49	250	$452,622.50
100-kilowatt generators	$10,000.00	5	$50,000.00
5,000-gallon steel holding tanks	$3,500.00	6	$21,000.00
1 horsepower pump system with pressure tanks, switches, and accessories	$3,000.00	6	$18,000.00
Latrine and/or shower	$39,700.00	16	$635,200.00
		Total:	$1,176,822.50

Base Camp Planning Factors

Table E-9. Maximum average estimated cost for SEA huts/SWA huts base camp

Item	Unit Cost	Quantity	Total Cost
Housing (SEA huts/SWA huts)	$5,215.98	250	$1,303,995.00
100-kilowatt generators	$10,000.00	5	$50,000.00
5,000-gallon steel holding tanks	$3,500.00	6	$21,000.00
1 horsepower pump system with pressure tanks, switches, and the like	$3,000.00	6	$18,000.00
Latrine and/or shower	$39,700.00	16	$635,200.00
		Total:	$2,028,195.00

Legend:
SEA   Southeast Asia
SWA   Southwest Asia

Table E-10. Sample planning factors for medical treatment facilities

Space	NSF	Notes
Medical	1,660	Based on the organic medical TOE staffing of a typical armor or infantry battalion (add 100 NSF per doctor/examination room); includes the functions listed in clinic requirements.
Dental	500	Minimal requirement for one dentist and one hygienist (add 115 NSF per dentist or DTR) two DTRs per dentist may be provided, depending on workload.
Holding	340	Minimal requirement for three-cot holding capacity (add 80 NSF per additional holding bed required).

Note. These sizes are NSF only and represent only a few of the common space planning factors for medical facilities. Consultation with a health facility planner or the Health Facility Planning Agency is imperative to ensure that proper space planning is completed in the Space and Equipment Planning System to determine appropriate solutions. A factor of 10 percent should be added for a gross estimate. In addition, a smooth transition for litters (ramping if necessary) should be added for entry into the main building with direct access to the trauma room.

Legend:
DTR   dental treatment room
NSF   net square feet
TOE   table of organization and equipment

E-44. Grossing factors are used to generally calculate total gross square footage from net square feet in all facilities. See table E-11.

Table E-11. Grossing factors

Condition	Percent of NSF
If a separate mechanical space is used	11 % of NSF
Circulation	35 % of NSF
Walls and partitions	12 % of NSF
Half areas	1.5 % of NSF
Total gross square footage	159.5 % of NSF

Legend:
NSF   net square feet
%   percent

E-45. For final planning, the exact number of physicians and dentists is obtained from the medical command. UFC 4-510-01 provides details regarding the development of heating, ventilation, and air conditioning requirements for military medical facilities.

Appendix E

UTILITIES FACTORS

E-46. Table E-12 provides planning factors to assist in evaluating options for developing estimates of some services provided by base camps. Table E-13 provides a visual perspective of power system cost considerations. Table E-14 describes power system planning. See appendix F for power system descriptions.

Table E-12. Base camp utilities planning factors

Item	Basic	Expanded	Enhanced
Water	10–13 gallons per person per day	30 gallons per person per day	50 gallons per person per day
Electricity	1.5 kilowatts per person	2.5 kilowatts per person	3.5 kilowatts per person
Wastewater	16 gallons per person per day	24 gallons per person per day	40 gallons per person per day
Solid waste	4 pounds per person per day	6 pounds per person per day	10 pounds per person per day

Table E-13. Power generation options versus costs

Type of Power Generation	Initial Cost	Operating Cost
Military	$	$
USG lease/military operators	$$	$
USG purchase/military operators	$$$	$
USG purchase/contractor operators	$$$	$$
LOGCAP owner/operators	$$	$$
Contract owner/operators	$	$$$
HN commercial	$	$

Legend:
$ low cost
$$ medium cost
$$$ high cost
HN host nation
LOGCAP Logistics Civil Augmentation Program
USG United States government

Table E-14. Power system planning considerations

Level of Services	Electrical Power System Attributes	Load/Situational Attributes	Operational Considerations
Platoon			
Basic	• Generator/power unit or power plant connected to isolated load(s) at user (low) voltage. (No attempt to consolidate power sources.)	Primary operational loads include— • Critical operations and command facilities. • Mission-essential communications. • Weapons/weapon systems. • Field feeding. • Mission-essential maintenance. • Mission-essential HVAC.	• Transportability and rapid setup to support mission-essential operations override other considerations. • Organic equipment operated by individual sections.

Table E-14. Power system planning considerations (continued)

Level of Services	Electrical Power System Attributes	Load/Situational Attributes	Operational Considerations
Platoon			
Expanded	• Power plants (capable of parallel operation) connected to consolidated loads, using PDISE (or similar) low voltage electrical distribution equipment. (Central Power Solution.)	Operational loads (in addition to the basic capabilities) include— • Supply and maintenance operations. • Laundry and shower facilities. • Dining facilities. • Life support areas (troop beddown, including HVAC).	• Deliberate system planning and/or coordination, is required. • Power plant(s) and PDISE equipment are organic to the utilities section and operated by utilities section personnel. • Individual unit organic power equipment is operationally controlled by utilities section.
Company			
Basic	• Power plants (capable of parallel operation) connected to consolidated loads, using PDISE (or similar) low-voltage electrical distribution equipment. (Central Power Solution)	Primary operational loads include— • Critical operations and command facilities • Mission-essential communications. • Weapons/weapon systems. • Supply and maintenance operations. • Laundry and shower facilities. • Dining Facilities. Life support areas (troop beddown, with HVAC).	• Deliberate system planning and/or coordination is required. • Power plant(s) and PDISE equipment organic to utilities section and operated by utilities section personnel. • Individual unit organic power equipment is operationally controlled by utilities section.
Expanded	• Deployable prime power system; medium voltage power generation and expedient distribution system; and secondary distribution centers (transformers) replace power plants. (Incorporate low voltage power systems into deployable prime power system)	Operational loads (in addition to those for basic capabilities shown above) include— • Water purification and distribution. • Ice production facilities. • MWR facilities. • Fitness centers. (Post/base exchange facilities [shoppettes and barber shops])	• Base camp master planning is required. • Deployable prime power platoon augments, the utilities section to install, operate, and maintain the deployable prime power system. • Power plants serve as redundant backup or can be reallocated to other sites.
Enhanced	• Expand deployable prime power plant as needed. • Expand and/or improve deployable prime power distribution system as needed. • Improved facilities can be designed to use waste heat from generators to preheat water for showers, laundry, to reduce fuel consumption.	Operational loads (in addition to those for basic and expanded shown above) include— • Expanded MWR facilities. • Post/base exchange vendors. • Theater maintenance and supply activities. • Improved/consolidated Dining facilities. • Improved shower and laundry facilities.	• Requires base camp master planning. • Deployable prime power platoon continues to augment utilities section for power system operation and maintenance. • Tent-based facilities transition to improved facilities.

Table E-14. Power system planning considerations (continued)

Level of Services	Electrical Power System Attributes	Load/Situational Attributes	Operational Considerations
Company			
Enhanced	• Expand deployable prime power plant as needed. • Expand and/or improve deployable prime power distribution system as needed. • Improved facilities can be designed to use waste heat from generators to preheat water for showers, laundry, and the like to reduce fuel consumption.	Operational loads (in addition to those for basic and expanded shown above) include— • Expanded MWR facilities. • Post/base exchange vendors. • Theater maintenance and supply activities. • Improved/consolidated Dining facilities. • Improved shower and laundry facilities.	• Base camp master planning, is required. • The deployable prime power platoon continues to augment the utilities section for power system operation and maintenance. • Tent-based facilities transition to improved facilities.
Battalion/Battalion Landing Team			
Basic	• Power plants (capable of parallel operation) connected to consolidated loads, utilizing PDISE (or similar) low voltage electrical distribution equipment. • Power Plants used initially, transitioning very quickly to deployable prime power system. • Deployable prime power system; medium voltage power generation and expedient distribution system; secondary distribution centers (transformers) replacing power plants.	Operational loads include— • Critical operations and command facilities • Mission-essential communications. • Weapons/weapon systems. • Supply and maintenance operations. • Laundry and shower facilities. • Dining facilities. • Life support areas (troop beddown including HVAC). • Water purification and distribution. • Ice production facilities. • MWR facilities. • Fitness centers. • Post/base exchange services (shoppettes and barbershops).	• Base camp master planning is required. • Power plant(s) and PDISE equipment is organic to the utilities section and operated by utilities detachment personnel. • Individual unit organic power equipment is operationally controlled by utilities detachment. • The deployable prime power platoon augments the utilities detachment to install, operate, and maintain deployable prime power system. • Power plants serve as redundant backup or can be reallocated to other sites.
Expanded	• Expand deployable prime power plant as needed. • Expand and/or improve deployable prime power distribution system as needed. • Improved facilities can be designed to utilize waste heat from generators to preheat water for showers, laundry, to reduce fuel consumption.	Operational loads (in addition to those for basic shown above) include— • Expanded MWR facilities. • Post/base exchange vendors. • Theater maintenance and supply activities. • Improved/consolidated dining facilities. • Improved shower and laundry facilities.	• Base camp master planning is required. • The deployable prime power platoon continues to augment utilities detachment for power system operation and maintenance. • Tent-based facilities transition to improved facilities.

Table E-14. Power system planning considerations (continued)

Level of Services	Electrical Power System Attributes	Load/Situational Attributes	Operational Considerations
Battalion/Battalion Landing Team			
Enhanced	<ul><li>Transition to sustained power system.</li><li>Retrograde deployable prime power equipment may be retrograded to Army pre-positioned stocks program.</li><li>Consider large-scale renewable energy.</li></ul>	<ul><li>All facilities connected to consolidated power system.</li><li>Expeditionary power systems (isolated generators and/or mini-grids) eliminated within the overall system.</li></ul>	<ul><li>Base camp master planning is required.</li><li>Long-term contracted support for operations and maintenance is required.</li><li>Utilities detachment personnel may serve as CORs.</li></ul>
Brigade/Regimental Combat Team and Larger			
Basic	<ul><li>Power plants (capable of parallel operation) connected to consolidated loads, utilizing PDISE (or similar) low-voltage electrical distribution equipment.</li><li>Power plants used initially, transitioning very quickly to deployable prime power system; combination of power units and power plants from organic elements.</li></ul>	Operational loads include—<ul><li>Critical operations and command facilities.</li><li>Mission-essential communications.</li><li>Weapons/weapon systems.</li><li>Supply and maintenance operations.</li><li>Laundry and shower facilities.</li><li>Dining facilities.</li><li>Life support areas (troop beddown including HVAC).</li><li>Water purification and distribution.</li><li>Ice production facilities.</li><li>MWR facilities.</li><li>Fitness centers.</li><li>Post/base exchange services (shoppettes and barbershops).</li></ul>	<ul><li>Base camp master planning is required.</li><li>Power plant(s) and PDISE equipment is organic to the utilities section and operated by utilities detachment personnel.</li><li>Individual unit organic power equipment operationally controlled by the utilities detachment.</li><li>The deployable prime power platoon augments the utilities detachment to install, operate, and maintain the deployable prime power system.</li><li>Power plants serve as redundant backup or can be reallocated to other sites.</li></ul>

Appendix E

Table E-14. Power system planning considerations (continued)

Level of Services	Electrical Power System Attributes	Load/Situational Attributes	Operational Considerations
Brigade/Regimental Combat Team and Larger			
Expanded	• Deployable prime power system; medium-voltage power generation and expedient distribution system; and secondary distribution centers (transformers) replace power plants. • The deployable prime power plant and distribution system is expanded and improved as needed. (Improved facilities can be designed to use waste heat from generators to preheat water for showers, laundry, and the like to reduce fuel consumption.)	Operational loads include— • Critical operations and command facilities • Mission-essential communications. • Weapons/weapon systems. • Supply and maintenance operations. • Improved/consolidated dining facilities. • Improved shower and laundry facilities. • Life support areas (troop beddown, including HVAC) • Water purification and distribution. • Ice production facilities. • MWR facilities. • Fitness centers. • Post/base exchange services (including barbershops). • Expanded MWR facilities. • Post/base exchange vendors. • Theater maintenance and supply activities.	• Base camp master planning is required. • Power plant(s) and PDISE equipment organic to utilities section and operated by utilities detachment personnel. • Individual unit organic power equipment operationally controlled by utilities detachment. • Deployable prime power platoon augments utilities detachment to install, operate, and maintain deployable prime power system. • Power plants serve as redundant backup or can be re-allocated to other sites.
Enhanced	• Transition to sustained power system. • May retrograde deployable prime power equipment may to the Army pre-positioned stocks program. • Consider large-scale renewable energy.	• Connect all facilities to the consolidated power system. • Eliminate expeditionary power systems (isolated generators and/or mini-grids) within an overall system, incorporate hybrid systems.	• Requires base camp master planning. • Long-term contracted support for operations and maintenance. • Utilities detachment personnel may serve as CORs.

Note:
¹ Includes refrigerated/frozen storage.

Legend:
COR — contracting officer's representative
HVAC — heating, ventilation, and cooling
MWR — morale, welfare, and recreation
PDISE — Power Distribution Illumination System Electrical

Other References

E-47. See ATP 3-37.34/MCTP 3-34C for protective structures planning factors for estimates of necessary materials, equipment, personnel, and total construction time. Additional techniques are presented in GTA 90-01-011. See UFC 4-010-01 for determining estimated standoff distances from various explosive charges. For additional potential engineer-specific base camp planning factors, see EP 1105-3-1. For further information about airfield standards, see UFC 3-260-01 and contact the International Civil Aviation Organization. Ammunition and explosives storage area planning factors may be found in DA Pamphlet 385-64. For medical facilities information, contact the Health Facility Planning Agency, Office of the Surgeon General. Contact the Army Geospatial Center for water detection response teams.

Appendix F

Base Camp Facilities and Infrastructure Design

Planning and design continue throughout the base camp life cycle as new or modified facilities and infrastructure are needed. During design, construction means, base camp standards, levels of base camp capabilities, on-site conditions, and adaptable, scalable designs are matched against facility and infrastructure requirements. The result is the production of detailed site designs, drawings, specifications, and special instructions needed for constructing facilities and infrastructure that make up the base camp. This appendix focuses on the two major tasks involved in facilities and infrastructure design. It describes the process and tools, such as the AFCS and geospatial information systems that base camp designers use to site-adapt standard facility designs.

DESIGN PROCESS

F-1. The following is a typical design process that can be used to design base camps. Design should be a collaborative process with planners, engineers, centers of excellence, and actual or potential base camp tenants. The design process consists of the following steps:
- Define life cycle requirements.
- Identify resources and constraints.
- Develop and conceptualize options.
- Evaluate options.
- Decide.
- Implement, assess, and adjust.

F-2. Design usually follows a top-down or bottom-up approach. A top-down approach begins with the purpose or function of a facility and works toward the identification of subcomponents and their interrelations, while the bottom-up approach starts with a set of given or implied components and works to arrange and link them to achieve desired results.

DEFINE LIFE CYCLE REQUIREMENTS

F-3. Defining life cycle requirements specifies the capabilities and attributes that the designed facility must have throughout its planned life cycle to achieve a specific purpose or function or to fulfill a need of the intended users. Defining requirements must often balance commander's guidance for QOL against a prescribed level of services, base camp standards, and available funding. Contingency construction designs are characterized by—
- Rapid construction or emplacement.
- Standardized, modular, and scalable designs.
- Maximized use of pre-positioned stocks and locally procured materials that meet military specifications.

F-4. For complex facilities, planners may prioritize requirements to facilitate evaluations and decision making. Typical important requirements should be identified as early as possible in the planning and design process and may include—
- Size (including life cycle variations).
- Purpose or mission.
- Unique tenant requirements.
- Expected duration.

Appendix F

- Level of services (including life cycle variations).
- Terrain selection.
- Command and staffing plan.
- Environmental factors.

IDENTIFY RESOURCES AND CONSTRAINTS

F-5. Resources which are the means for constructing a facility or structure, include labor, materials, equipment, and funds. The availability of materials depends on the local market, access to other markets (which may be determined by the military and political situation), transportation assets, and available funding. The labor for base camp construction may be supplied by military units and contractors. Each labor source has certain strengths and weaknesses based on equipment, training, and experience. If a certain labor source, (such as HN workers) is more prominent than others, it may be beneficial to select designs that can be executed with the capabilities and skills of that labor source. Certain designs may not be supportable based on the available labor pool or the availability of materials. The results of planning should provide most of the information needed for this step.

F-6. A constraint dictates an action or inaction, thus restricting how something can be done. Design is constrained by such things as—

- Base camp standards (facility allowances and construction standards).
- The availability of construction resources (labor, equipment, materials, and money).
- The quality and availability of indigenous material.
- Terrain and weather effects on construction methods and materials.
- The amount of land area that is available and useable.
- Funding limits.
- AT/force protection requirements.
- HN agreements.
- Environmental considerations and effects.
- Operational timelines.
- Civil considerations and effects on the local economy, resources, and population.
- Characteristics, the availability, and the reliability of local commercial power.

DEVELOP AND CONCEPTUALIZE OPTIONS

F-7. Planners consider established techniques, methods, and local practices to simplify planning, material, and labor requirements. Planners should also consider using reachback. The designs completed at this point are called concept or preliminary designs.

F-8. Base camp designers use two different techniques to design: requirements-based and capabilities-based. Requirements-based design looks primarily considers fulfilling all identified requirements and determining what resources are needed in terms of manpower, equipment, materials, funding, and time available. This is often referred to as an unconstrained approach. Capability-based design primarily considers the capabilities of the force to perform construction based on manpower, equipment, materials, funding, and time available.

EVALUATE OPTIONS

F-9. Planners determine the advantages and disadvantages of each option based on evaluation criteria which address factors that affect success and those that can cause failure. The base camp principles may be used for evaluation criteria and life cycle cost analysis.

F-10. Preliminary estimates for material and construction requirements developed during planning are updated and completed in more detail during design. A detailed cost estimate is developed to allow a cost comparison of one or more concept designs. A detailed cost estimate is also developed to analyze the engineering tradeoffs made to complete the detailed designs. Assumptions may be required to complete a life cycle economic analysis of design options. The operational commander should consider this life cycle cost estimate when selecting from the design options that form the basing strategy and base camp construction directives. The commander may direct design tradeoffs and resource-constrained designs.

DECIDE

F-11. Planners determine the best option based on their evaluation. After the evaluation is complete, they recommend it for approval by the designated approving authority.

IMPLEMENT, ASSESS, AND ADJUST

F-12. Planners make the necessary coordination and prepare orders or directives needed for implementing the approved option. The detailed designs of specific facilities may be developed at a higher headquarters or by the constructing units. The designated approving authority ensures that detailed designs conform to approved standards and the master plan, and approves the designs before contracting or construction begin. Site-adapted designs are generally approved by the headquarters that completed the concept designs. The constructing unit and base camp staff have the necessary engineering expertise (or obtain it through reachback), automated design tools, access to standard designs, and network capability to share, archive, and print construction documents. Planners and the constructing unit assess the progress and compare forecasted outcomes with actual events to determine the overall effectiveness. Based on the assessment, adjustments are made or new options are developed to achieve the desired results. Lessons learned and recommended improvements to standard designs and theater adaptations are captured to facilitate design modifications, and base camp records and as-built designs are updated and maintained to facilitate future construction and transfer or closure.

GENERAL DESIGN CONSIDERATIONS

F-13. The UFC system provides planning, design, construction, operations, maintenance, sustainment, restoration, and modernization criteria for all DOD components. UFCs will be used for all Service projects and work for other customers, where appropriate. Most UFC are written with long-duration standards in mind and apply only to permanent installations or bases. Base camp planners and designers will find them useful for contingency construction as well. There are few UFCs specific to contingency construction and these should be consulted as appropriate (see UFC 1-201-01 and UFC 1-201-02). In addition to construction standards established by the CCDR for the AOR, construction outside of the United States may also be governed by DOD guidance, status-of-forces agreements, HN-funded construction agreements and in some instances, bilateral infrastructure agreements. Planners and designers must ensure compliance with the more stringent standards, as applicable. UFC are effective upon issuance and are distributed only in electronic media from the following source: Whole Building Design Guide. See UFC 1-200-01 for information on general building requirements.

F-14. Incorporating the building styles often found on permanent installations and bases, such as stucco exterior walls, multiple interior walls, indoor plumbing, and large windows, increases resource requirements and construction timelines and is generally not used in contingency construction. However, certain structures–such as perimeter fencing; ECPs; segmented, compartmentalized information facilities, and airfield runways–are normally constructed to the same standards that are applied in permanent installations and bases, regardless of location. These designs are available in various engineering TMs or through reachback.

F-15. Developing right-size facilities and utility infrastructure can quickly become a false economy. Aside from getting the right amount of land in the right location, utility infrastructure (such as electric, water, communication, and sewer lines) can be the tightest physical chokepoint of base camp operational surges. Installing deliberately oversized utility runs, including easements, offers the greatest flexibility in evolving, uncertain base camps. Base camp life cycle requirements are assessed to estimate maximum and minimum requirements to enable an effective, scalable design.

Appendix F

ANITITERRORISM/FORCE PROTECTION

F-16. Minimum contingency requirements normally include hardened walls and roofs to protect occupants and sized to primarily accommodate the personnel. UFC 4-010-01 establishes DOD AT standards for buildings. Implementation of these minimum standards is mandatory for all expeditionary structures that meet the occupancy criteria for inhabited or primary gathering buildings or billeting. Facilities should be designed to resist attack through material selection, a minimization of the number of doors and windows, and the orientation of openings to minimize exposure. Overhead blast protection designs can be incorporated into all contingency construction facilities and are available as a retrofit for existing structures. The most common design is a layered structure, with one layer used to detonate incoming munitions and a second layer absorbing the blast concussion and shrapnel.

F-17. The AFCS incorporates limited AT/force protection requirements into its designs. AT/force protection designs fall into two main categories: isolation and hardening. For isolation, most designs will need to be augmented. One alternative is using soil-filled containers to isolate the facility. For hardening, CMU walls can be hardened with reinforced concrete up to the expected blast height. A number of other blast mitigation products have been developed by Service component laboratories, such as ERDC. Many are presented in ATP 3-37.34/MCTP 3-34C and GTA 90-01-011. See ATP 3-37.2 for more information on determining threats, assessing vulnerabilities, and integrating AT measures within operations.

FIRE PROTECTION

F-18. Austere environments often lack adequate water and maintenance resources to support modern fire suppression systems. More combustible materials tend to be used in the construction, of temporary structures. Consequently, fire events can result in the rapid loss of facilities and can spread quickly to other structures. An effective fire protection plan is critical to the safety of personnel, facilities, and equipment. Fire protection must be included in the design of base camps. This includes tent and building spacing, means of egress, wiring standards, the use of flame-retardant materials, fire-fighting vehicle access, the availability of a water supply, and fire protection and HAZMAT spill response equipment. See TM 3-34.30 and UFC 3-600-01 for more information.

SAFETY

F-19. Risk management is initiated during planning and continues throughout the base camp life cycle. Designers work together with safety specialists in mitigating the risks associated with any hazards.

F-20. Design influences safety during construction. Some designs and the associated construction methods may be more difficult, especially when unskilled labor is used, and inherently more dangerous. Designers must ensure that the complexity of designs is reasonable and justifiable based on the construction means available and/or that the means for enforcing safety and mitigating risks during construction are achievable. HN laborers and contractors may require additional oversight to ensure adherence to expected construction and safety standards.

F-21. Any specifications in component configurations, materials, and construction tasks that are essential for achieving the quality and safety features of the design must be clearly articulated to the constructing unit and become part of the overall quality assurance and surveillance plan. Any incorrect design decisions, changes desired by the facility user, or material substitutions based on availability may require the reevaluation of designs.

STRUCTURAL INTEGRITY

F-22. The risk from structural collapse increases greatly with the transition from the use of tents to hardened structures and with the use of existing facilities. Although contingency construction standards are generally conservative in order to address a wide range of loads in different environments, the structural integrity and conditions of an existing structure can vary greatly based on HN construction standards and quality of construction and the effects of battle damage. Existing structures may have little resistance to seismic activity, abnormal weather, or impact loads. The base camp engineer or professional engineer must oversee the allowable use of existing structures. A structural analysis and materials evaluation must be completed before any protection measures are affixed to an existing structure. The use of reachback may be necessary where field reconnaissance is inadequate or technical capability is unavailable.

F-23. The base camp engineer oversees any repairs, modifications, or expansions of any existing building to ensure they conform to established policies and standards. This requires a complete structural assessment before occupying an existing structure. Construction variances with structural components that deviate from the Service standards and material substitutions for structural members with standard designs require a structural assessment and compliance with UFCs.

CONSTRUCTION MATERIALS

F-24. Although using locally-procured construction materials offers many advantages, there are several factors to consider including, that—
- **Standard sizes may be different.** Dimensional lumber is often cut to different standard lengths in foreign countries. For example, some countries may cut lumber in meters rather than feet.
- **Quality may be substandard.** Lumber, concrete, and asphalt are three examples of foreign construction materials that are typically not consistent with U.S. standards.
- **Military operations may drive up prices in the local market.** Sudden spikes (or perceived increases) in demand may result in profiteering from local suppliers.

F-25. Hazardous construction materials such as asbestos and lead-based paint may be discovered in existing structures during renovation. This should be a focus of reconnaissance efforts and infrastructure assessments prior to occupying any existing structures.

DRAINAGE SYSTEM

F-26. The planning and design of base camp drainage systems are conducted by higher headquarters design engineers and the constructing unit. The drainage system includes the overall drainage plan, area drainage structures, individual facility drainage structures, and temporary construction drainage. Siting of base camps and individual facilities can have major effects on required drainage structures and their associated cost in terms of materials and construction effort. Inadequate drainage is the most common cause of road and airfield failure. Data on local drainage conditions for initial planning may be obtained from maps and aerial reconnaissance and then confirmed with on-site ground reconnaissance and information from local inhabitants. See TM 3-34.48-1, TM 3-34.48-2, AFPAM 10-219, Volume 7, for discussion of drainage system design.

F-27. Following are some drainage consideration tips:
- Site base camps and individual facilities in locations that minimize required drainage structures and their associated cost in terms of materials and construction effort.
- Evaluate the natural and existing drainage features, expected rainfall or snowmelt, and protection of natural drainage channels.
- Avoid constructing facilities in areas with high water tables.
- Develop the drainage system and temporary drainage features in phases to ensure uninterrupted construction.
- Perform continuous maintenance on the drainage system.

MULTIPURPOSE BUILDINGS

F-28. Buildings or areas are needed for base camp services and support, administration, and command functions. Depending on the scope of the base camp, many of these functions can be collocated within unit headquarters and administration buildings to minimize space requirements. MWR activities may require separate buildings or areas, such as sport fields. Depending on the amount of traffic, landing areas may also serve as sports fields to maximize the use of limited space. Areas designated for expansion may also be used as temporary sports fields.

Appendix F

BILLETING

F-29. When new construction is authorized for billeting, several design options may be available including tents, prefabricated trailers, wooden SEA huts/SWA huts, or concrete/masonry construction. One variation of the SEA huts/SWA huts is the Davison SEA hut, which is combination six SEA huts in order to save on materials. In certain climatic environments, facilities where personnel are billeted or work require heating, ventilation, and air conditioning systems. See ATP 3-34.40/MCTP 3-40D for more information.

TOILET AND SHOWER FACILITIES

F-30. Toilet and shower facilities should be lighted and heated, with efficient fixtures and equipped with hot and cold water. Sanitary wallboard is the preferred wall covering for latrines. Sheetrock, if used, must be waterproofed by some means. The female-to-male facility ratio is based on the actual percentage of the sexes on a base camp at the current time or anticipated for the near future. A ratio of 1:20 is generally used for determining the number of shower heads and toilets needed in relation to the number of personnel of actual or future occupation. Electrical outlets must be grounded and protected from contact with wet conditions near toilets and showers.

DINING FACILITIES

F-31. Tactical field kitchens provide initial food service operations for base camps. Once the CCDR determines that forces will remain in the theater for an extended period of time, base camps may transition to food service support through garrison type dining facilities. These may consist simply of a tent in which personnel eat in or a structure that resembles a permanent facility on an installation. Food service personnel generally prepare meals using mobile kitchen trailers that can feed up to 350 people or containerized, trailer-mounted systems that can feed up to 600 people. Contractor-operated dining facilities, on the other hand, can be quite large and require extensive cooking and food storage facilities (including refrigeration requirements, food waste containment, and grease traps). Regardless of the size, dining facility operations require large quantities of water for cooking and cleanup. While units in the field may establish gray water soakage pits for dining facility wastewater, larger base camps require other waste management options. See engineer doctrine for more information on waste management operations. See FC 4-722-01F for information on Service-specific dining facilities construction.

HAZARDOUS MATERIAL/HAZARDOUS WASTE STORAGE FACILITIES

F-32. Controlling and managing HAZMAT/HW protect residents of the base camp and the environment. Most units use large quantities of HAZMAT, such as ammunition, fuels, paints, batteries, pesticides, and solvents. Often, these compounds contain acids, metals, and other toxins. This is one of the first environmental protection issues that should be addressed at base camps. Design specifications or standard facility designs should be provided for base camps of all sizes. See ATP 4-35.1 for information on secondary-blast mitigation for ammunition and explosives storage, TM 38-410 for information on HAZMAT management, UFC 4-451-10N for information on designing HW accumulation points, and engineer doctrine for information on spill prevention and cleanup and secondary containment.

MOTOR POOLS AND VEHICLE PARKING AREAS

F-33. Requirements for motor pool facilities and vehicle parking areas must be identified early during base camp planning and synchronized with the traffic flow patterns and AT/force protection requirements as part of land use planning of the base camp. Requirements can include areas for conducting maintenance (enclosed and exterior maintenance pads), administrative functions, vehicle washing; and POL storage. Parking areas should be constructed using well-graded, compacted rock and soil with an engineered slope and drainage to minimize weather effects and improve the safety and longevity of the parking area. Important considerations include scalability to facilitate expansions, drainage, storm water management, and environmental requirements such as oil-water separators.

Transportation Infrastructure

F-34. Base camp standards, operational requirements, the availability of construction equipment and materials, and soil composition (soil type, drainage characteristics, and moisture content) influence road design and construction. Designers need to account for material availability when planning for expeditionary construction. Naturally occurring construction materials, such as rock, may be scarce or of poor quality. Portland cement of any type may not be available or may be cost-prohibitive.

F-35. Soils may require stabilization to increase strength and durability or to prevent erosion and dust generation. Regardless of the purpose for stabilization, the desired result is the creation of a soil material or soil system that will remain in place under the design use conditions for the design life of the project. Matting and sand grid are expedient methods for stabilizing loose soils, such as sand, for unsurfaced road construction. Geotextiles and other geosynthetics are primarily used to reinforce weak subgrades, maintain the separation of soil layers, and control drainage through the road design. Geosynthetics are the primary means of waterproofing soils when grading, compaction, and drainage efforts are insufficient. See TM 3-34.48-1 and TM 3-34.48-2 for more information on expedient surfacing methods and TM 3-34.64/MCRP 3-40D.7 for more information on soil characteristics and predicting soil behavior under varying conditions.

F-36. Planners balance technical engineering design and construction considerations with the desired degree of permanence to generate options that are optimized for effectiveness, while, maintaining as much efficiency as possible. Roads are designed and built with the understanding that future improvements will be necessary to sustain continued use and to accommodate higher volumes of traffic as base camp populations increase. Based on anticipated future needs and the characteristics of the expected traffic, plans are developed to progressively improve roads as time and resources become available and the situation allows. Requirements for road maintenance and upgrades are incorporated into the base camp master plan. Planners must ensure that roads do not interfere with the routing of underground utility lines.

F-37. The base camp commander/BOS-I is responsible for traffic management to ensure safe and efficient movement throughout the base camp. Military police play an important role in traffic control by helping to identify requirements for traffic control posts and implementing the necessary measures to enforce speed limits. See FM 3-39 and TM 3-11.42/MCTP 10-10G/NTTP 3-11.36/AFTTP 3-2.83 for more information.

F-38. An airfield or landing zone may be required based on the base camp purpose or functional requirements. Most base camps need a minimum of a landing zone to facilitate resupply operations and casualty evacuation. See TM 3-34.48-1 and TM 3-34.48-2 for information on designing airfields and landing zones.

Waste Management Facilities

F-39. Deployed forces can generate significant amounts of waste. Roughly 80 percent of all water used on base camps for purposes other than human consumption ends up as wastewater and requires treatment and disposal. The base camp commander/BOS-I is responsible for waste management that involves collecting waste at its point of generation and transporting it from collection points to treatment, disposal, or recycling facilities to provide a healthy and sanitary environment for base camp residents. Waste management facilities are designed to handle each category of waste: wastewater (gray and black water), solid waste, hazardous and special waste, and medical waste. The measures used to treat and/or dispose of waste vary according to the base camp population, level of services, and availability of existing facilities, contracted support, base camp location, terrain effects, and civilian and environmental considerations. More permanent collection, treatment, and disposal facilities may be possible at expanded or enhanced base camps.

F-40. Solid waste sources consist of food, packaging materials, and other sources of trash such as office supplies. Liquid waste can be broadly divided into categories; gray water, black water, and industrial waste water. Sources of gray water include waste products from showers, sinks, and laundry. Black water is composed of waste product from latrines, and kitchens. An example of industrial waste water is be reverse osmosis (RO) reject; a by-product from water RO purification. Designs for wastewater lagoons and scalable wastewater treatment package plants are available as listed in AFCS. Extensive general engineering support or contractors are required to build and maintain such systems.

Appendix F

F-41. The waste streams generated on a base camp place a significant demand on a unit's resources. Reusable water containers, such as government-issued personal hydration systems, could be used instead of disposable plastic bottles to reduce waste and the added logistical strain associated with back-haul.

F-42. Waste storage and disposal areas should be positioned downwind, downstream, and downhill from all other areas of the camp. Additionally, consideration should be given to positioning waste collection points outside of the base camp to minimize physical security risks and health hazards. See TM 3-34.56 MCIP 3-40G.2i for more detailed information on waste management.

MEDICAL FACILITIES

F-43. The provision of medical care to address the needs of the wounded, injured, and ill is a critical service to be considered during the planning and construction of a base camp, regardless of its size. The planning and management of a health facility is a complex task requiring interaction between multiple trades and disciplines that are influenced by the following factors: the role of medical care to be provided (roles 1, 2, and 3 medical care in-theater and role 4 in the continental United States or other safe haven), length of the sustainment period, availability of evacuation assets, operational theater maturity level, and distance to medical treatment facilities outside the theater.

F-44. As a theater or contingency matures, the need to establish or improve physical plants and ensure an environment of care that is more supportive of clinical and operational requirements increases. Facilities should provide the right medical capability at the appropriate location. Continuous improvements in quality and safety result in cleaner and more durable facilities with reliable power, water, lighting, climate control, public addresses and patient care systems. The underlying driver is the inherent need to upgrade facilities to support ever-increasing equipment modernization, greater electrical loads, improved utilities reliability, and greater patient safety (such as electrical safety, life safety, and other code requirements).

F-45. The level of medical support and the type of facilities vary and should be taken into consideration when planning base camps. In a deployed setting, the specifics range from aid stations, dental and medical clinics to the combat support hospitals. The lowest-level facility is the battalion aid station, which is a Role 1 medical treatment facility organic to maneuver battalions and most field units. This forward medical treatment facility enables reception, triage, emergency treatment and disposition of wounded, injured, or ill individuals performed by medical personnel. Role 2 medical care is provided by the medical company (area support) operating at echelons above brigade level and the medical company (brigade support battalion) providing coverage at brigade and below, (as mission requirements progress) may be augmented to transition into a more robust solution such as a medical clinic where medical treatment activities provided include advanced trauma management and tactical combat casualty care and ambulatory services. These facilities may also perform nontherapeutic activities related to the health of the personnel served (physical examinations; immunizations; and medical administration including PVNTMED services). Additionally, clinics may be equipped with a holding capability for the observation of patients who are awaiting transfer to a hospital or the care of patients who cannot be seen as outpatients without requiring hospitalization.

F-46. Another important factor to consider regarding the planning of medical treatment facilities is the availability of existing deployable solutions. The Deployable Medical Systems facility solution is organic to military medical units and used across DOD. These solutions are mobile/deployable, modular in nature (thus scalable), capable of being relocated, in existence (no immediate procurement action required), and coordinated and outfitted with the associated medical equipment. These facility solutions do have limitations in durability and survivability; are generally intended to operate on dual-voltage/frequency systems (110/220 volt and 50/60 hertz); and may provide the capability needed to satisfy requirements. In the absence of Deployable Medical Systems, existing medical treatment facilities or structures that are easily adaptable for such use should be considered. See ATP 4-02.1 for additional information on health facility planning and management.

WATER PRODUCTION AND DISTRIBUTION SYSTEMS

F-47. Components of a water production and distribution system may include water sources, water purification, well drilling, storage, and water distribution.

WATER SOURCES

F-48. Planners estimate base camp water requirements based on the base camp population and level of capability and on sources of water. Commanders and planners must consider the impact of base camp water sources and usage on local aquifers and water sources used by local populations. Sources of water include local municipal water utilities; water generation through water purification systems and wells; water distribution through storage tanks and pipes to facilities, water trailers and blivets; and bottled water. Water purification and well drilling present more sustainable alternatives to the use of bottled water. Plastic water bottles significantly add to the generated solid waste at a base camp and present disposal challenges when local recycling is unavailable. Some operations generate effluent water that can be reused after minimal treatment. Recycling water from showers, sinks, laundries, washracks, and other nonpotable water sources is a considerable conservation mechanism and should be practiced whenever feasible. This can bring the base camp into a more sustainable posture.

WATER PURIFICATION

F-49. The production of bulk water is often accomplished by water purification—generally through the use of water purification systems that can be operated by troops and contractors. Water purification units require adequate operational space, and they must be within 5 kilometers of a water source. The water purification process also generates wastewater that must be managed. See UFC 3-230-03, ATP 4-44/MCRP 3-40D.14, and TC 4-02.3 for more information.

WELL DRILLING

F-50. Well drilling can be performed by specialized engineer units and contractors. Planners must determine the availability of well drilling capabilities and the viability of drilling based on a hydrogeological analysis of the area. Initial information on the hydrogeology of an area is available through geospatial engineering channels or reachback to USACE.

F-51. This analysis should also incorporate a test well-drilling program. Drilled wells may be integrated into a water distribution system within the base camp, or water may go into storage tanks or bladders for distribution by vehicles. There is normally a low chemical or biological threat of contamination, and there is not usually a large seasonal variation in groundwater quantity. After PVNTMED personnel test of groundwater and approve a groundwater source, treatment is not usually required; however, chlorination is recommended. See TM 3-34.49/NAVFAC P-1065/AFMAN 32-1072 for more information.

F-52. The Army Geospatial Center provides water resources information to the military community under the authority of DODD 4705.1E.

F-53. If the treatment of ground water is required, allowing the ground water to sit exposed to the open air for a short duration before the purification process is recommended. Some compounds found in ground water oxidize when exposed to air, then precipitate out of solution, putting less strain on the RO water treatment unit systems.

F-54. The U.S. Army Reserve Components maintain the bulk of the well-drilling capability within DA. The U.S. Air Force RED HORSE Squadrons and U.S. Navy also have well-drilling capabilities. Wells can be drilled to a depth of 1,500 feet.

BULK WATER STORAGE

F-55. Existing military water storage systems were designed to support tactical expeditionary forces for a short amount of time. These systems are fabric-based and are susceptible to degradation from prolonged exposure to ultraviolet light and other harsh conditions. See TM 10-4320-303-13 for more details on maintenance and use of military water storage systems.

F-56. There are commercial alternative systems for large volume water storage when military water storage systems are not available. Advertised products are engineered for free-standing or anchored installation on sites with minimal or no preparation. Tanks can be fitted with a floating cover if desired. If this or a similar commercial solution or product is pursued, the use of a cover is recommended. A number of smaller tanks

Appendix F

that can meet the equivalent capacity is encouraged, if a tank becomes a single point of failure, the other tanks can maintain the supply.

F-57. All product water or potable water must maintain a chlorine residual. Numerous factors affect how quickly chlorine residual decays, such as temperature, the duration that water is stored, and the volume of water stored. Stored water requires more circulation and additional chlorination over a long period of time to maintain drinking water status.

WATER DISTRIBUTION SYSTEM

F-58. Gravity and direct pressure are the two major forms of water distribution. If a direct-pressure distribution system is required, a properly sized water pump that provides the required flow and pressure is installed at the base of the tower. The configuration of the water distribution system is determined primarily by the size and location of water demands, road patterns, location of treatment and storage facilities, and topography. Two patterns of main distribution systems commonly used are the branching or dead-end pattern and the gridiron pattern with a looped feeder or central feeder system. In general, mains should be routed to minimize the length of service connections. Mains should not be located under paved or heavily traveled areas and should be separated from other utilities to ensure the safety of potable water supplies and minimize the interference of utilities maintenance. Water supply storage and distribution systems are susceptible to sabotage. Safeguarding the supply must be considered during the design process. See TM 3-34.70/MCRP 3-40D.5 and UFC 3-230-01 for more information.

ELECTRICAL POWER GENERATION AND DISTRIBUTION SYSTEMS

F-59. Base camp power systems are as varied as the sizes, locations, and missions of the base camps they support. Identifying the power requirements of a site and integrating a generation and distribution network that is appropriate for the size and duration of a base camp are significant parts of effective base camp master planning. Reliable and sustainable power systems are composed of modular and scalable components ranging from Soldier-operated tactical power systems, to hybrid and deployable prime power systems operated by specially-trained engineer Soldiers, to a sustained power system operated and maintained by civilian personnel and may include local commercial power. During base camp development and evolution, all three types of power systems may exist concurrently on an individual base camp. Figure F-1 depicts a typical base camp power life cycle.

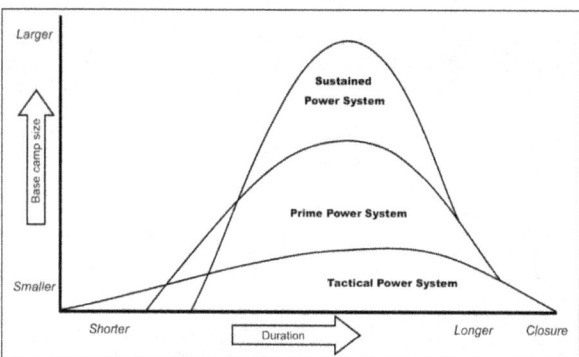

Figure F-1. Base camp power life cycle

F-60. The base camp power life cycle consists of the following systems:
- **Tactical power system.** This is a highly mobile system composed of modified table of organization and equipment-authorized generators (typically 300-kilowatt and below). It uses an electrical distribution system, such as the Power Distribution and Illumination System Electrical or the Mobile Electric Power Distribution System Replacement. The initial power system may be augmented with commercial, off-the-shelf generators or electrical distribution equipment operated and maintained by military personnel.
- **Prime power system.** This is a deployable system composed of large, rapidly deployable generators (typically 500-kilowatt and larger) that can be consolidated to operate as a power plant. It uses a medium-voltage electrical distribution system capable of distributing power over the entire footprint (greater than 5 miles, if necessary) of the base camp. The deployable prime power system is scalable and able to supply reliable, utility-grade power needed for base camp support and services and tenant unit operational requirements. The consolidation of electrical loads and reduced number of generators required yields improved fuel economy and overall reduction in the O&M costs. The deployable prime power system may employ generators to provide redundant backup power. It may also be augmented with commercial off-the-shelf generators or electrical distribution equipment that are operated and maintained by military personnel.
- **Sustained Power System.** As the situation changes and resources become available, a commander may direct a base camp to transition toward longer-term, sustained operations. Responsibility for the operational control, sustainment, and maintenance of the existing initial power system(s) and deployable prime power system(s) is transferred to civilian, contracted, or HN personnel. Life-cycle equipment replacement and further expansion of the power system creates a site-specific sustained power system that is typically composed of fixed, commercial generators (or utility power, if available) and commercial electrical equipment that continues to be operated and maintained by civilian or contracted personnel.

POWER SYSTEM EFFICIENCY

F-61. An efficient electrical power system minimizes the sustainment/logistics support (fuel) required to meet the electrical demand. Two primary considerations to optimize power system efficiency are to maximize the efficiency of the power production and distribution system (supply side management) and to minimize the consumption of electrical power (demand side management). Planners must properly size generators (to meet electrical demand) and specify fuel-efficient generators with electronic fuel-management systems. The electrical distribution system must be designed as compactly as practical to minimize electrical losses, and electrical loads must be consolidated to ensure generators are operated with maximum efficiency. Power demand can be reduced by using energy-efficient equipment (especially generators and ECUs), improving the thermal efficiency of structures (with tent linings, building insulation, solar shades), and incorporating other energy conservation measures (lighting timers, occupancy sensors, programmable or timer-controlled thermostats).

POWER SYSTEM DESIGN CONSIDERATIONS

F-62. Base camp master planning must address and manage current demand, and future power system growth. Base camp power systems contain three elements: a power source (such as generator, power plant, and batteries), a distribution system (power distribution panels, power cables, transformers) that delivers the power, and the load or consumer (such as air conditioners, lighting, and communication equipment). It is imperative that the CCDR decide whether the primary base camp power system will be constructed to U.S. standards (120/208 volts at 60 hertz) or to local standards (230/400 volts at 50 hertz, or other standard). However, there may be situations where U.S. and local standard power are required. In these cases, separate power systems or additional equipment to convert frequency and transform voltage to the appropriate standard may be required. Power distribution systems for basic-capability; level base camps may have surface-laid power cables (and covered with a protective shield when located in high-traffic areas), while expanded- and enhanced-capability level base camp power distribution systems will likely have buried and/or overhead power cables. Regardless of the voltage and frequency standard, is followed, or method of power distribution, appropriate safety measures must be implemented to prevent damage to cables and to reduce

Appendix F

electrocution hazards. AutoDISE is an Army computer modeling program to assist master planners designing power distribution, illumination system, and electrical systems.

GENERATOR PLACEMENT

F-63. Generators should be placed as close as possible to the point of demand, for spot generation, without disrupting other activities, such as meetings or sleep; to minimize the materials needed for the distribution system: and to avoid voltage drops that may impair equipment function. They must also be positioned to allow for easy service and maintenance, particularly refueling. Sufficient space must be allotted for the placement of fuel bladders, safety equipment, and fire extinguishers. Planners will also need to account for fuel resupply at the bladder (a means to refuel the bladder). Generators must be located away from buildings, walls, or other obstructions that may impair cooling to multiple linked systems. Typically, at least 5 feet of clear space is required—at least 10 feet between generators. Prevailing wind direction may be considered to aid in generator cooling. Sandbags, partitions, and barriers may be placed around generators to reduce noise, as long as they do not obstruct cooling air flow.

GENERATOR PROTECTION

F-64. Generators must be protected against attacks, unauthorized access, and the elements. Protection measures may include overhead roofs, protective walls or berms, and secondary containment measures for fuel leaks and spills. The use of protective walls or berms also helps to reduce noise pollution, as long as they do not obstruct cooling air flow.

POWER SYSTEM RESOURCES

F-65. Larger camps with expanded and enhanced capabilities typically rely on deployable prime power that uses large generator power plants with distribution systems provided by deployable prime power units or contracted services. This transition away from spot generation to power distribution systems and commercially produced power typically results in cost savings and improved fuel use efficiency. Modular base camp life support sets, such as Force Provider and Harvest Falcon, include organic generation capability that is generally sufficient for its internal components that are designed for a specific number of occupants. Reliable commercial grid power should be used whenever possible, with the appropriate amount of backup power generation available when needed for critical facilities. See TM 3-34.45 for more information on deployable prime power planning considerations.

F-66. When base camp power requirements exceed a unit's organic capabilities, there are several resources that may provide additional power system capacity. The Army's Force Provider system is a base camp life support module, which is configured to 150-people scalable up to a 3,600-person configurations and comes with a power generation system. For larger or longer-duration operations, the Force Provider Prime Power Kit enables the transition and connection to the deployable prime power system. See ATP 4-45 for more information. The Air Force's Basic Expeditionary Airfield Resources System is a deployable airfield operations package, which is complete with low-voltage and high-voltage power systems. If funding is available, additional power system equipment and support may be obtained from indefinite duration, and indefinite quantity, contract resources, such as the Army Materiel Command LOGCAP contract. Additional power system equipment is available through the General Services Administration or local contract sources, and may require proper system redesign to ensure equipment is utilized safely and efficiently.

ALTERNATIVE/RENEWABLE ENERGY

F-67. Planners should leverage renewable energy sources such as solar, waste-to-energy, and wind whenever possible to help make base camps more sustainable. The proper employment of renewable energy sources requires foresight during base camp master planning. Energy conservation measures include renewable energy systems (photovoltaic arrays, solar collectors for power and hot water) that can be reliably integrated into smart base camp micro-grids without harming the grid stability or degrading the output of the renewable source. A combination of spot generation with renewable energy sources is generally not recommended. Plans should account for regional wind patterns and features, such as mountains and buildings that may block the solar resource when allocating space for renewable energy systems on base camps.

POWER SYSTEM DESIGN RESOURCES

F-68. Most units are not trained to establish power distribution networks. Units may request support from specialized units such as the 249th Engineer Battalion (Prime Power), or the USACE Forward Engineer Support Team or contract for design support. The risks for electrocution and fire are substantial concerns with electrical systems, and the design must address whether the power system is constructed to U.S. or HN specifications. The adopted codes must be consistent with the voltage and the building materials available and the construction methods used. See the Whole Building Design Guide for more information.

This page intentionally left blank.

Appendix G
Reachback

Reachback is the process of obtaining products, services, applications, forces, or equipment, or materiel from organizations that are not forward-deployed. The reachback objective is to assist base camp planners and operators through specialized technical and communications assistance not normally found in deployed units. Deployed units and personnel are linked to subject matter experts within USACE, the U.S. Army Engineer School and Service equivalents, theater engineer commands, USACE districts and divisions worldwide, other government agencies, academia, and private industry.

USACE REACHBACK OPERATIONS CENTER

G-1. The United States Army Corps of Engineers Reachback Operations Center (UROC) at the U.S. Army ERDC provides a reachback engineering capability to support contingencies across the full operational- and natural-disaster spectrum. The UROC rapidly leverages extensive USACE resources to deployed forces or those requiring specialized assistance. Common topics of RFIs include—
- Bridge assessment.
- Airfield design and repair.
- Dam breach and hydrologic analysis.
- Trafficability (on- or off-road).
- Geophysical environmental analysis.
- Climate information and analysis.
- Force protection/survivability.
- Bomb damage assessment.
- Critical infrastructure assessment.
- Base camp development.
- Engineering, master planning, and facilities.
- Construction and mapping.

G-2. The UROC is shown in figure G-1, page G-2. UROC can also be reached by telephone at 601-634-2439 or Defense Switched Network (DSN) 312-446-2539. E-mail requests can be sent to <uroc@usace.army.mil>.

Appendix G

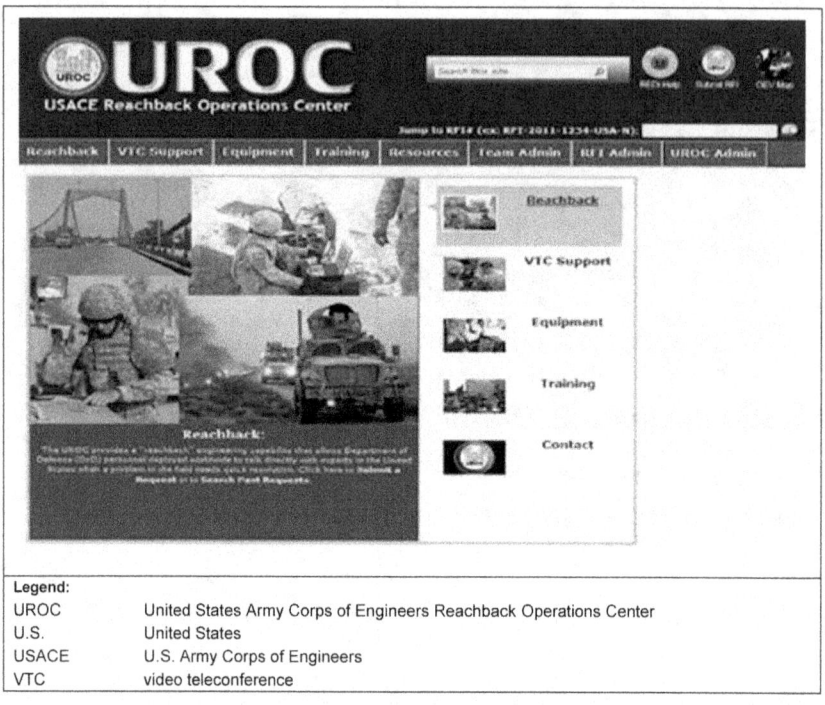

Legend:	
UROC	United States Army Corps of Engineers Reachback Operations Center
U.S.	United States
USACE	U.S. Army Corps of Engineers
VTC	video teleconference

Figure G-1. UROC Web site

REACHBACK ENGINEER DATA INTEGRATION

G-3. The USACE REDi System provides a common database; a robust user interface; and fully integrated mapping tools for receiving, managing, tracking and archiving all data and engineering reachback support conducted through UROC. The UROC REDi portal allows users to submit RFIs; receive updates and track the status of RFIs; search the historical RFI database; and request support for other UROC capabilities such as reachback equipment, training, and video teleconferencing support. REDi is also used by the UROC staff as a tool for managing and documenting support for all UROC customers and supported elements. To date, UROC has developed custom portals within the REDi architecture for the USACE G-2, USACE Field Force Engineering Program, USACE Transatlantic Division, CENTCOM J4 Engineers, USARPAC DCSENG, Army Facilities Component System, and others. See figure G-2. Custom features, tools, permissions and site layout are provided for each portal while advantage is taken of a single, common, overall hardware and software system architecture. A common access card is needed for site access.

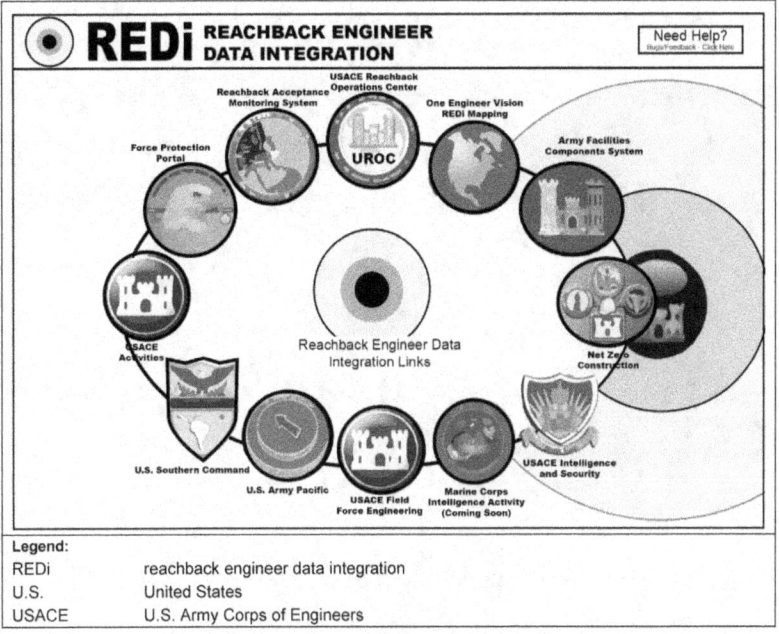

Figure G-2. REDi home page

NAVAL FACILITIES ENGINEERING COMMAND REACHBACK

G-4. NAVFAC contingency engineering provides a reachback capability through NAVFAC Atlantic and NAVFAC Pacific. NAVFAC forwards RFIs to the appropriate supporting organization or subject matter expert. The NAVFAC Reachback Information Portal is common access card enabled for portal access.

AIR FORCE CIVIL ENGINEER CENTER REACHBACK

G-5. Headquarters, AFCEC, operates the Reachback Center (RBC), which provides rapid response to questions from civil engineers in the field, worldwide. The RBC is dedicated to assisting DOD civil engineers in the execution of their mission across the full operational spectrum. The RBC can be reached by phone at 850-283-6995 (toll-free 888-232-3721) or DSN 312-523-6995. E-mail requests can be sent to AFCEC.RBC@us.af.mil.

This page intentionally left blank.

Appendix H
Base Camp Communications Support

This appendix discusses communication network requirements for base camps. It addresses the requirements to support the operational needs of tenant and transient units in company, battalion, and battalion/battalion landing team base camps. It describes the roles and responsibilities of the base camp commander/BOS-I, the G-6/S-6, and supporting units providing communications support. Although base camps may initially be established by other Services, this manual focuses on required communications when base camps are established or expanded to support land component operations. This appendix pertains primarily to the Army and the Marine Corps. Providing long-term base communications support is not a traditional mission requirement for the Marine Corps; however it is fully capable of doing so with appropriate augmentation.

Note. For the purpose of this appendix, communication refers specifically to electronic communications and technical networks.

COMMUNICATIONS SUPPORT CONSIDERATIONS

H-1. Communications support requirements for base camps are based on the functions, services, and support necessary for base camp operations and the operational needs of tenant and transient units. These requirements are very similar to the services provided by the network enterprise center at permanent installations. On larger base camps, the installation of a commercial communications infrastructure is necessary to replace and free up tactical resources and provide for a more robust, longer-duration network. This includes planning for environmental control requirements, power generation, and network redundancy.

H-2. Planning, installing, operating, maintaining, and managing the network architecture of a base camp, as part of the DOD Information Network, requires the right kind of equipment and personnel with the necessary technical skills. This includes NETOPS, network management, information dissemination management, and information assurance. Most of these capabilities are organic at the BCT/RCT level and below and may require augmentation.

DETERMINING REQUIREMENTS

H-3. The communications support element determines the capabilities that are required for each base camp during each phase of the operation. Requirements exceeding the organic capabilities of the supporting communications unit are fulfilled through augmentation that is requested through the appropriate channels. Considerations include base camp size, life cycle, tenants, and support of transient units, as applicable. These capabilities can be grouped within the following areas:
- Local- and wide-area networks.
- Secure and nonsecure telecommunications services.
- Contracted services and support.

LOCAL- AND WIDE-AREA NETWORKS

H-4. The local area network (LAN) is a computer-based network covering a small geographic area or group of facilities (such as a base camp). A wide-area network (WAN) covers a broader area or multiple base camps in an AO. LANs and WANs operate at sensitive, unclassified, and classified levels. LAN and WAN services may include electronic e-mail, data sharing, access to the Web, desktop publishing, graphics applications,

Appendix H

voice-over Internet protocol (VoIP), and video teleconferencing over Nonsecure Internet Protocol Router Network (NIPRNET) and SECRET Internet Protocol Router Network (SIPRNET).

H-5. Each theater may have a multinational network, such as the Combined Enterprise Regional Information Exchange System. These multinational networks are the backbone in supporting information-sharing requirements between unified action and interorganizational partners. Extended services may be needed for remote or nonstandard users, such as multinational forces, HNs, non-state actors, nongovernmental organizations, and others that may have unique communications needs.

SECURE AND NONSECURE TELECOMMUNICATION SERVICES

H-6. The DSN and Defense Red Switched Network (DRSN) are part of the Defense Information System Network, which provides nonsecure and secure telecommunications services, including point-to-point switch voice, data, imagery, and video services to all of DOD. There are three separate networks for handling—NIPRNET, SIPRNET, and the Joint Worldwide Intelligence Communications System. DSN can be extended to joint and multinational subscribers as the technology matures. DRSN provides secure voice and video teleconferencing. Marine Corps units provide DRSN only to the highest commands (Marine expeditionary force commander).

H-7. The Joint Worldwide Intelligence Communications System provides continuous, classified, compartmented, point-to-point, or multipoint information exchange involving voice, text, graphics, data, and video teleconference up to the top secret, sensitive, compartmented information level through a system of interconnected computer networks. It is an intelligence counterpart to the SIPRNET for worldwide secure multimedia intelligence communications.

CONTRACTED SERVICES AND SUPPORT

H-8. Contractors are often needed for maintenance and communications support. During planning, the G-6/S-6 identifies the need for, and duration of, contracted support. The G-6/S-6 coordinates with the G-4/S-4, and other staff members, as required, for the protection and oversight of contractors. The base camp commander/BOS-I is responsible for the protection of contractors living and working on base camps, while contractor coordination cells within higher headquarters are responsible for accounting and arranging deployment support for CAAF.

H-9. It is crucial that the units requiring contractor support be identified early in the planning process. Contractor support for sustained base camp operations after turn-over to the theater tactical signal brigade (TTSB)/MAGTF G-6 differ from contractor support provided to the TTSB/MAGTF G-6 and must be considered during planning.

H-10. Program managers may contract support into the life cycle of communication systems. The contract may require the contracted companies to provide system support in the form of maintenance, new equipment training, or system configuration and management. Contractor support for communications-electronics systems is coordinated by the G-6/S-6 or signal/communications unit commanders. Considerations for contracted support should include—

- Fire protection and life support, as required by contractual obligations similar to other tenants.
- The identification of requirements/restrictions for using local civilians, civilian internees, and detainees in sustainment/logistics support operations.
- Battlefield procurement and contracting.
- Coordination with the staff judge advocate on legal aspects of contracting.
- Coordination with financial management staff personnel on the financial aspects of contracting.
- Real property control.

COMMUNICATIONS SUPPORT CONFIGURATIONS

H-11. Deployed forces expect base camps to provide many of the same types of network and information services that were provided at home station. It is important to include the communications support that is required during the base camp master planning. Planners should incorporate the base camp principles of scalability, sustainability, standardization, and survivability in designing the network architecture. The information systems and networks must be of sufficient scale, capacity, reach, and reliability to support evolving operational requirements or a reduction in operations, which is more important during transfers and closures of base camps. During base camp design, units can provide proposed designs for the planned timeframes, depending on the assigned units, personnel, equipment, and material. The base camp communications support element should take into consideration the types of services required by users and tenant units organic capabilities. These considerations include how many classified and unclassified voice and data terminals and DSN and DRSN capabilities are required for operational use. The primary design of a base camp communications support infrastructure must consider the resources and personnel necessary to meet uncertain requirements in often austere environments. Additional base support activities such as mass notification and warning systems, MWR, post/base exchange services, and QOL will affect the overall plan for communications support.

H-12. The specific mission and the operational requirements of tenant units of the base camp are considered in developing the communications plan. This plan is composed of specific, modular components that are primarily focused on duration-related operational needs and the expected life cycle of the base camp. Each base camp is unique in how it is arranged and the number and types of tenant and transient units and organizations that it supports. Structural and security issues are critical considerations in developing the communications support structure for each base camp. Design accounts for more robust facilities and the expected end-life use for the HN, including considerations for what may be left behind for use by the HN. System efficiencies should be implemented as applicable to optimize communications support, especially on longer-duration base camps.

H-13. The designated unit or base camp must be prepared to support the signal/communications support elements with rations, billets, POL, and sustainment/logistics. These life support functions vary based on the number of communications support personnel needed to sustain communications and communications maintenance efforts. The base camp commander/BOS-I is responsible for connectivity and infrastructure support, excluding computers, telephones and other assets. Communication between tenant units and the BOC is imperative to coordinate changes in projected support requirements.

H-14. Base camp communications support may have non-U.S., nongovernmental base camp tenants who may require access to the communication services provided. Network security requirements should be implemented using approved security measures.

H-15. The placement of communications support equipment, antenna interference prevention, and protection of communication lines must be coordinated early during base camp planning to optimize communications and protect all equipment from interference, damage, and exploitation. Cable runs must be protected from vehicular traffic.

H-16. The appropriate communications unit provides connectivity to the DOD Information Network for company, battalion, or battalion/battalion landing team base camp configurations as required. Units need to provide their own data systems or identify their communications requirements during the initial planning. The category-5 cable, fiber, and data connectivity need to be established for each supported unit before it deploys. Expeditionary signal battalions (ESBs) are a pooled Army resource to support units with communications assets.

COMPANY BASE CAMP CONFIGURATION

H-17. Four or more battalion size headquarters that lack, or have limited, organic communications assets, may reside on a company base camp. Communications requirements vary greatly based on the type of tenant units, the number and types of subscribers, and the required bandwidth. An example of possible support for this size base camp is an Army ESB, and ESB platoon, or a Marine Corps communications platoon/element/detachment that provides voice, VoIP, data, imagery, or video services. See figure H-1, page H-4. Subscriber requirements are mission- and equipment-availability dependent. If a company base camp communications requirements closely matches the capability of an ESB platoon, this will likely be the element in support. If the requirements are significantly less or more, then a different set of communications support assets may be required. See Army and Marine Corps doctrine for information on the organization and capabilities of signal/communications units.

Appendix H

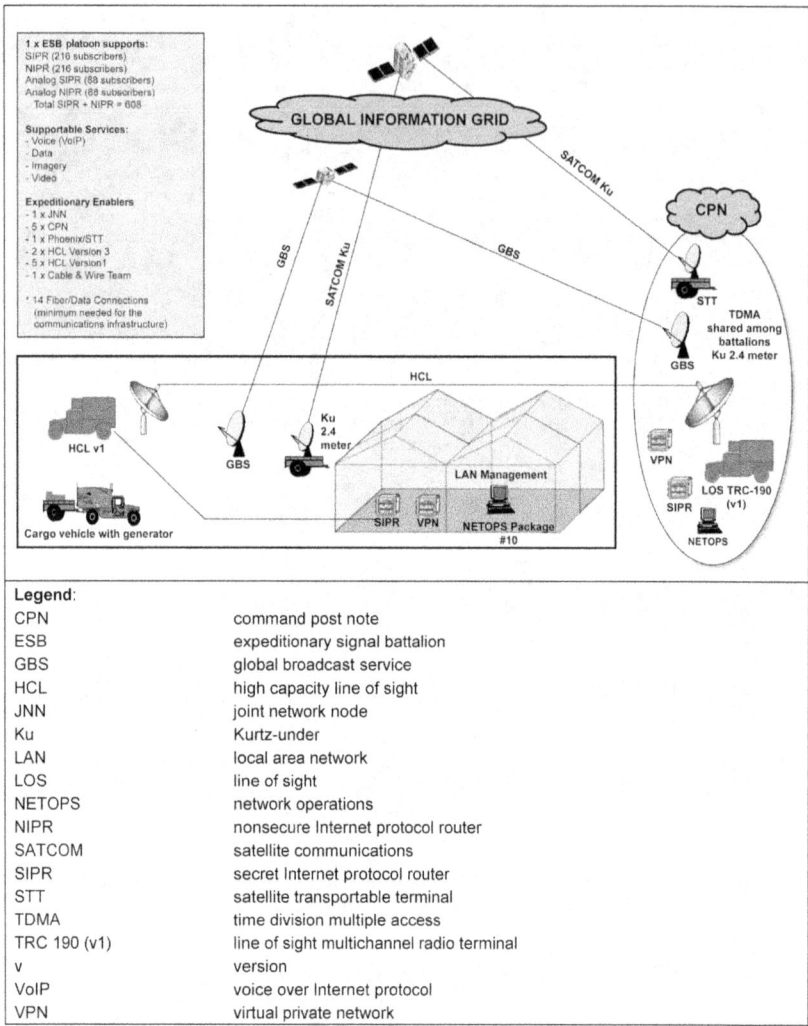

Figure H-1. Example of Army signal support configuration for a company base camp

BRIGADE COMBAT TEAM BASE CAMP CONFIGURATION

H-18. BCT base camps typically contain at least one brigade/regiment size unit including two or more battalion size headquarters that have limited organic communications capabilities. Communications requirements vary greatly based on the types of tenant units, number and type of subscribers, and required bandwidth. An example of possible support for this size base camp is an Army ESB company or a Marine

Corps communications detachment/company/squadron to provide voice, VoIP, data, imagery, and video services. See figure H-2, page H-6. If medium-size base camp communications requirements closely match the capability of an ESB/communications company, this will likely be the element in support. If the requirements are significantly less or more, then a different set of communications support assets may be required.

SUPPORT AREA BASE CAMP CONFIGURATION

H-19. A support area base camp can typically contain one division or corps headquarters and several brigade/regiment size units that often have limited organic communications assets. For the Marine Corps, a support area base camp supports a Marine expeditionary force and all tenant commands assigned. Each supported command is directed to provide organic communications personnel and equipment, as required, to better support the mission. An example of a support area base camp is an Army ESB or a Marine Corps communications battalion to provide voice, VoIP, data, imagery, and video services. See figure H-3, page H-7. If a support area base camp has communications requirements that closely match the capability of an Army ESB or a Marine Corps communications battalion, this will likely be the element in support. If the requirements are significantly less or more, then a different set of communications support assets may be required.

MARINE AIR-GROUND TASK FORCE COMMUNICATIONS

H-20. The typical equipment used by the Marine Corps to connect MAGTF units is shown in figure H-4, page H-8. The senior MAGTF element at the base or base camp is responsible for validating and granting all tenant unit communication requests. The MAGTF, the G-6/S-6 uses only the communication equipment necessary to ensure mission success based on the mission and size of the MAGTF. The communication plan will be articulated in annex K of the OPORD. This plan defines the roles and responsibilities for all communications units in the MAGTF.

LONG-TERM REQUIREMENTS

H-21. The transition from tactical communications support to a long-term, commercial network communications infrastructure requires the following actions:
- Installing network redundancies.
- Procuring fixed telecommunications hardware.
- Establishing cyber support facilities.
- Installing fee-for-service commercial television to troop billets (as required).
- Providing possible contractor support for infrastructure sustainment or improvement.
- Establishing long-term telecommunications.

Appendix H

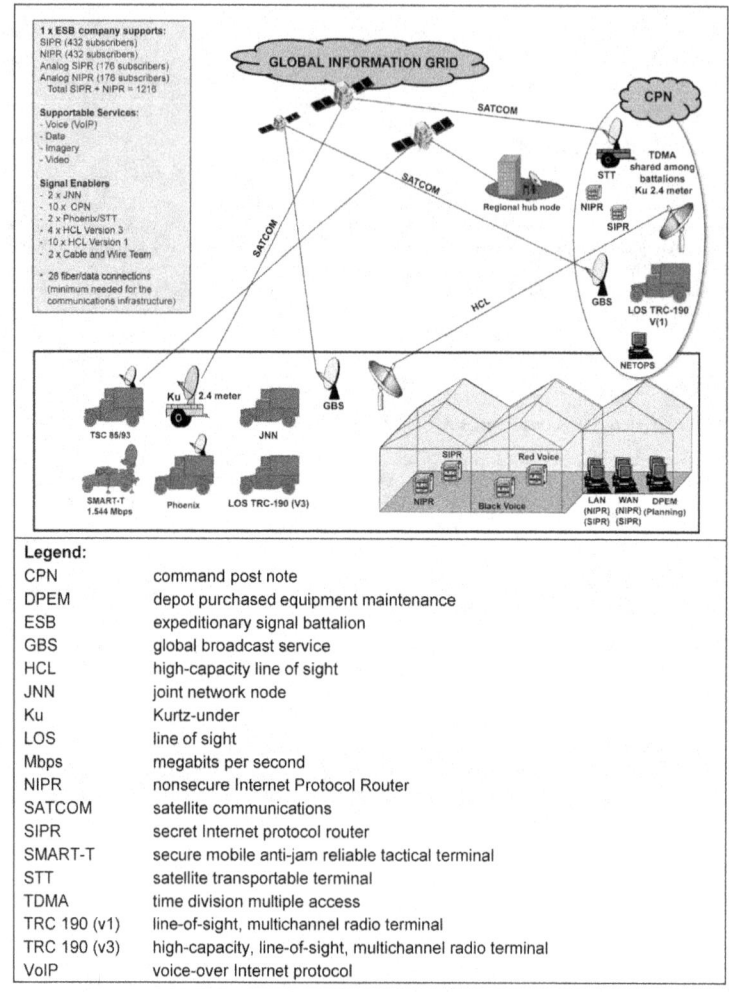

Figure H-2. Example of Army signal support configuration for a battalion/battalion landing team base camp

Base Camp Communications Support

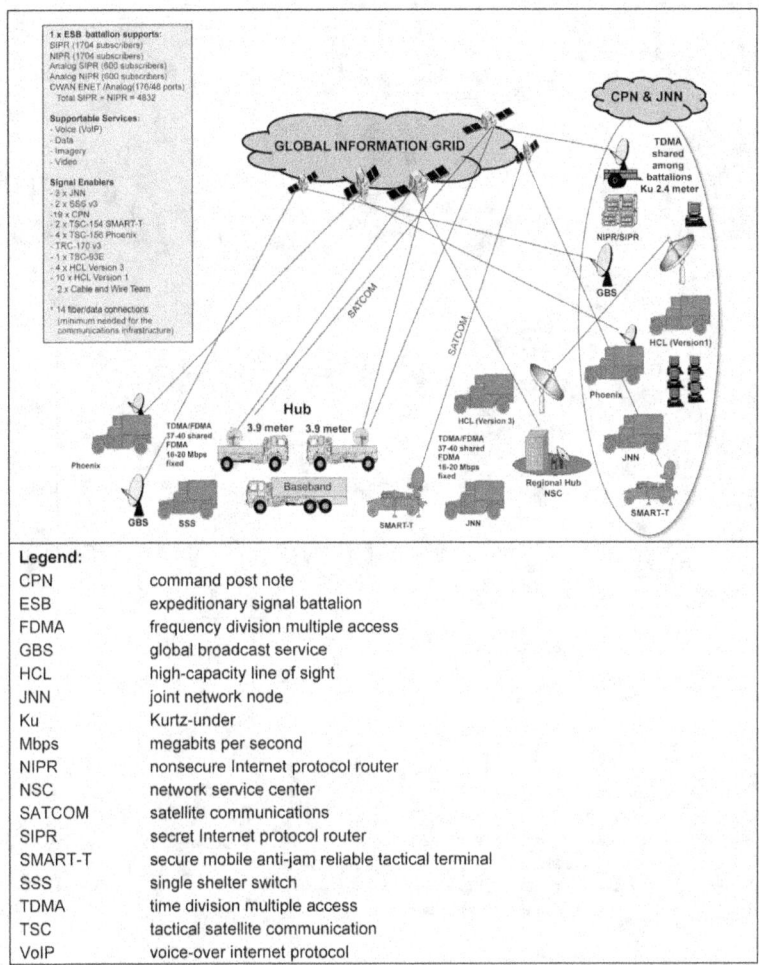

Legend:
CPN	command post note
ESB	expeditionary signal battalion
FDMA	frequency division multiple access
GBS	global broadcast service
HCL	high-capacity line of sight
JNN	joint network node
Ku	Kurtz-under
Mbps	megabits per second
NIPR	nonsecure Internet protocol router
NSC	network service center
SATCOM	satellite communications
SIPR	secret Internet protocol router
SMART-T	secure mobile anti-jam reliable tactical terminal
SSS	single shelter switch
TDMA	time division multiple access
TSC	tactical satellite communication
VoIP	voice-over internet protocol

Figure H-3. Example of Army signal support configuration for a support area base camp

Appendix H

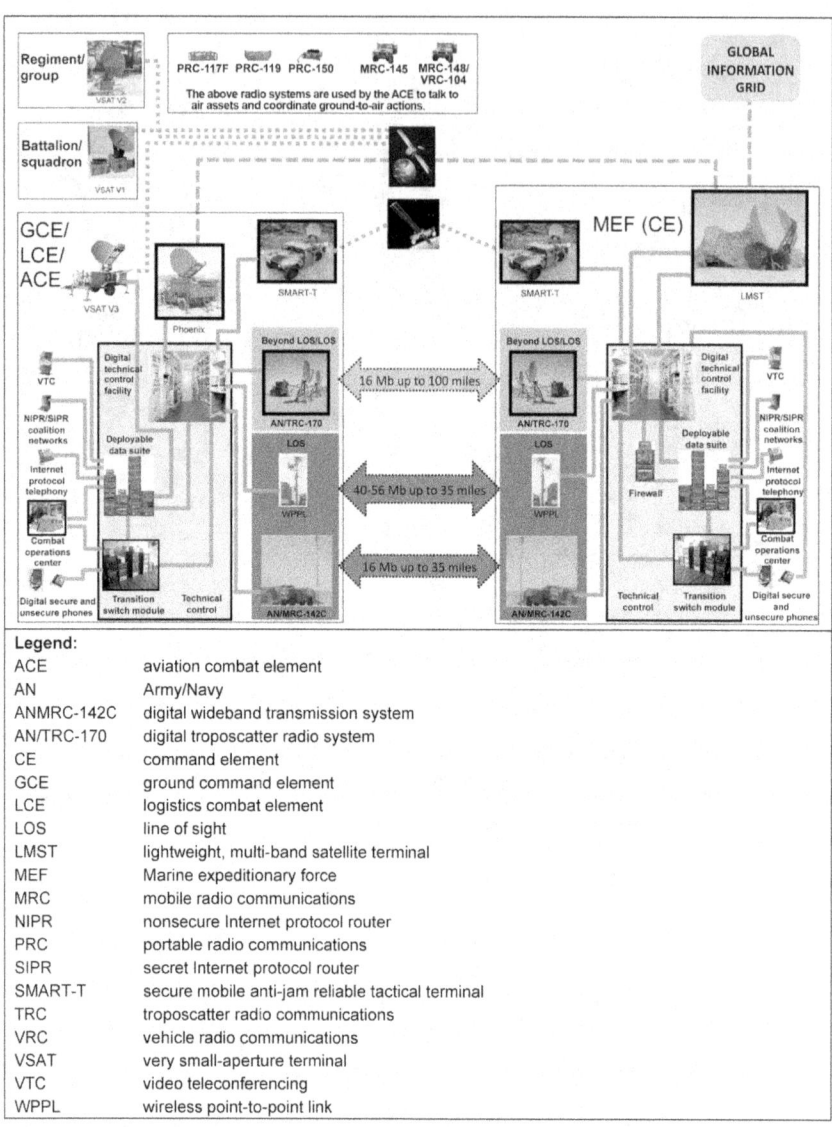

Legend:
ACE	aviation combat element
AN	Army/Navy
ANMRC-142C	digital wideband transmission system
AN/TRC-170	digital troposcatter radio system
CE	command element
GCE	ground command element
LCE	logistics combat element
LOS	line of sight
LMST	lightweight, multi-band satellite terminal
MEF	Marine expeditionary force
MRC	mobile radio communications
NIPR	nonsecure Internet protocol router
PRC	portable radio communications
SIPR	secret Internet protocol router
SMART-T	secure mobile anti-jam reliable tactical terminal
TRC	troposcatter radio communications
VRC	vehicle radio communications
VSAT	very small-aperture terminal
VTC	video teleconferencing
WPPL	wireless point-to-point link

Figure H-4. MAGTF communication architecture overview

H-22. The G-6/S-6 evaluates different designs and the flexibilities of those designs in adapting to situational changes in the mission and operation. The communications system directorate of a joint staff (J-6)/G-6 addresses potential threats related to the designs and operational needs. The J-6/G-6 also ensures that base camp communications are interoperable with unified action/interorganizational partners. Communications support planning and design must also include mechanisms to enable global sourcing of equipment, materials, and contract support to enhance the expeditionary capabilities of base camps.

H-23. Army ESBs and Marine Corps communications units assigned to base camp operations are adaptable and tailored to meet the needs of the JFC. The planning and design are modular and scalable to adapt to the size of any operational element, and they have the flexibility to be adjusted based on operations or conditions. Required equipment for all possible scenarios and conditions is identified. Equipment and repair parts that are not organic to units or that require long lead times to acquire or deploy may be pre-positioned and maintained by the appropriate unified action/interorganizational partners or contracting agencies (such as the Army Materiel Command).

H-24. The G-6/S-6 plans and coordinates several tasks with the responsible base camp commander/BOS-I, including—
- A means for performing reachback for operational issues.
- Site reconnaissance (engineering, infrastructure, environmental, health, and safety).
- Master plan writing and modification.
- Required closure planning and coordination.
- Construction management and oversight for initial construction, expansion, and deconstruction.
- Design modifications.
- Contract management.
- Scaling of facilities to account for surges or consolidation.

TRANSFERS AND CLOSURES

H-25. Communications support should be scaled to that required during base camp transfers and closures. This should be a planned effort between the communications support element, the base camp commander/BOS-I, and tenant units to ensure that the communications support is adequate during downsizing operations.

H-26. The G-6/S-6 is responsible for assuring the communications portion of base camp master plans is being implemented, modified, updated, and maintained to provide continuity for unit rotations. The G-3/S-3 and supporting Army ESB and Marine Corps communications units are responsible for developing the closure plans and reports. They are also responsible for anticipating and planning any required site actions (such as deconstruction and environmental cleanup) for their respective base camps.

H-27. In support of transfers and closures, and deconstruction actions are likely the last actions performed at any given base camp. Deconstruction actions include the removal or decommissioning of structures, systems, and facilities. Deconstruction actions ensure that the base camp is returned to a state agreeable to the JFC and the HN, whether that includes the transfer of facilities or reduction to preoccupancy conditions.

H-28. Base camp tenant units should expect to experience a gradual reduction in network capabilities as base camps prepare for transfer or closure. The base camp commander/BOS-I determines the mission-essential communications requirements needed for sustaining protection (security and defense) and seamless flow of communications between command nodes during the execution of transfer and closure actions.

H-29. During the last few days before transfer or closure, the base camp is operated and managed by a skeleton crew that is capable of conducting final closeout or transfer inspections and handling any remaining property transfer actions. In some instances, base camps may be closed and kept in a warm shut-down condition in anticipation of some future use. In these cases, the staffs may consist of a combination of military and civilian personnel.

Appendix H

ROLES AND RESPONSIBILITES

H-30. Planning the allocation of communications assets is based on each base camp and its specific mission. Required roles and responsibilities, enable tenant and transient units to effectively conduct and sustain operations to ensure connectivity within each base camp.

THEATER ARMY/MARINE AIR-GROUND TASK FORCE ASSISTANT CHIEF OF STAFF FOR COMMUNICATIONS

H-31. The theater/MAGTF G-6 is responsible for developing a theater signal/communications support plan for base camps that supports the theater basing strategy. This plan describes the communications support to base camps for each phase of the operation, (if the operation is phased,) and addresses the necessary actions needed for expanding and reducing base camps as the operation progresses. The theater/MAGTF G-6 provides the necessary details in plans and orders to facilitate execution, monitors the situation, and recommends adjustments to facilitate the efficient and effective use of signal/communications assets.

H-32. The theater/MAGTF G-6 is the senior signal/communications officer who exercises staff oversight of the communications support and information network and has the level of experience to anticipate and implement the necessary adjustments to meet changes in the situation. The G-6 derives this network from the CCDR and is empowered to use all available communications equipment and personnel to accomplish the mission. The G-6 is accountable for all network transport, network services, and the viability of information systems throughout the theater. The G-6 controls these network assets through NETOPS functions and uses technical service orders, such as a fragmentary order, to implement changes.

H-33. The theater/MAGTF G-6's responsibilities encompass all the management, control and defense capabilities for the theater network. The G-6 is organized and resourced to provide NETOPS support. The G-6 uses NETOPS functions to synchronize disparate networks into one information network as a part of the LandWarNet. The NETOPS functions performed in the subordinate support brigades, Marine expeditionary brigades/forces, and BCT/RCTs provide a second echelon of NETOPS management that the G-6 coordinates as part of the greater NETOPS plan.

ASSISTANT CHIEF OF STAFF FOR COMMUNICATIONS/SIGNAL STAFF OFFICER

H-34. The G-6/S-6 is the principal staff officer for all matters concerning communications and networks. The G-6/S-6 has technical oversight responsibility over the command's information networks, to include training and readiness oversight of subordinate communications units. The G-6/S-6 is responsible for providing planning guidance to subordinate communications units to execute the command, control, communications, and computer plan in support of the commander's intent. In executing the commander's intent, the G-6/S-6 directs any technical changes to the network. To make physical moves to communications equipment, the G-6/S-6 recommends fragmentary orders that will direct such movement to the G-3/S-3. The G-6/S-6 is responsible for advising the commander, staff, and subordinate commanders on command, control, communications, and computer-operational matters (staff responsibilities, technical guidance, and training and readiness oversight).

STAFF RESPONSIBILITIES

H-35. G-6/S-6 staff responsibilities include the following:

- Prepare, maintain, and update command, control, communications, and computer operations estimates, plans, and orders. Such orders often cause configuration management changes across multiple units.
- Monitor and make recommendations on the technical aspects of command, control, communications, and computer operations.
- Advise the commander, staff, and subordinate commanders on command, control, communications, and computer operations and network priorities (for example, changing bandwidth allocation).
- Direct technical changes to all portions of the base camp network via the technical service order process.

- Develop, produce, change, update, and distribute signal operating instructions.
- Prepare and publish command, control, communications, and computer operation SOPs.
- Coordinate, plan, and manage the electromagnetic spectrum operational environment within the AO.
- Plan and coordinate with higher and lower headquarters regarding information systems upgrade, replacement, elimination, and integration.
- Work with the intelligence, operations, and knowledge management staff officers to coordinate, plan, and direct all information assurance activities and command, control, communications, and computer operations vulnerability and risk assessments.
- Coordinate with other staff members and a variety of external agencies to develop the information and communications plans, manage the information network, obtain required services, and support mission requirements.
- Confirm and validate user IRs in direct response to the tactical mission.
- Establish command, control, communications, and computer policies and procedures for the use and management of information tools and resources.
- Coordinate cable routing and physical protection.

TECHNICAL AUTHORITY RESPONSIBILITIES

H-36. G-6/S-6 coordination and technical oversight responsibilities include the following:
- Provide subordinate communications units with direction and guidance during the preparation of network plans and diagrams establishing WANs, including business and intelligence WANs.
- Plan and integrate information systems and equipment in response to unit task organization and reorganization.
- Plan and direct all NETOPS activities within the AO in coordination with the Service component command and joint task force.
- Use the NETOPS and security center as the eyes and ears to the network, and leverage the tools provided by the NETOPS and security center to manage and reconfigure the network as warranted.
- Manage and control the use of information network capabilities and network services throughout the AO.
- Manage radio frequency allocations and assignments, and provide spectrum management.
- Manage the production of user directories and listings.
- Provide communications support to information operations.
- Recommend locations for base camps based on communications support considerations.
- Coordinate with the plans officer on the availability of commercial information systems and services for military use.
- Manage all communications support interfaces with unified action/interorganizational partners, including HN support interfaces.
- Coordinate, update, and disseminate command frequency lists.
- Manage communications protocols, and coordinate user interfaces of defense information systems.
- Advise the commander on support requirements versus support assets available.
- Coordinate external support requirements for supported units.
- Synchronize support requirements to ensure that they support current and future operations.
- Plan and monitor support operations, and make necessary adjustments to ensure that support requirements are met.

TRAINING AND READINESS RESPONSIBILITIES

H-37. G-6/S-6 training and readiness responsibilities include the following:
- Determine the commands, efficiency, economy, morale, and readiness.
- Ensure the development of required skills to all communications personnel within the AO.

Appendix H

- Identify requirements and manage the distribution of communications personnel in coordination with the personnel staff officer.
- Monitor and provide oversight for information dissemination to adjust to changing warfighting function priorities and control measures within the AO in coordination with the G-3/S-3.
- Ensure that automation systems and administration procedures for all automation hardware and software being used are compliant with the DOD Information Network procedures and standards or Service specifications.
- Ensure that assigned communications units are trained to support missions and tasks during home station training events and deployments in coordination with the parent unit commander.

COMMUNICATIONS-ELECTRONICS MAINTENANCE

H-38. The overarching principle of replace forward/fix rear remains unchanged. Modular organizations continue to build on the two-level maintenance system, composed of field maintenance and sustainment maintenance. The two-level maintenance system is one that essentially combines unit and direct support levels of maintenance (field maintenance) and general support and depot levels (sustainment maintenance). Field maintenance involves on-system tasks, normally performed by assets internal to a unit, which return systems to a mission-capable status. At field level, all functions are focused on replacing and returning to the user. The goal is to reduce repair cycle times by providing capabilities as far forward on the battlefield as possible, minimizing reliance on parts distribution, visibility, and replacement. Sustainment maintenance involves off-system tasks that are performed primarily in support of the supply system (repair and return to supply). There are no fixed repair time guidelines for performing field or sustainment repair. In the modular organization, maintenance procedures and doctrinal methods are changed to gain greater effectiveness and efficiencies. Regional support centers are maintained by contract for the repair of evacuated equipment, and serves as a repository for spare parts, within their respective regions.

MAINTENANCE ON COMMUNICATIONS-ELECTRONICS SYSTEMS

H-39. At the battalion level, the S-6 is responsible for field level maintenance on communications-electronics systems. The S-6 works in conjunction with the S-4 and the supporting forward support company to provide a comprehensive maintenance plan that is then incorporated into the unit maintenance SOP. This effort ensures that clearly understood procedures in place for a positive maintenance posture. The S-6 must also coordinate with the S-4 for contractor maintenance support as necessary. Consider the necessary maintenance coordination prior to deployment to ensure the integration of equipment organic to or allocated for use by, the G-6/S-6. At the brigade/regimental level, the S-6 is responsible for monitoring the status and sustaining brigade/regiment networks. The brigade/regiment S-6, working closely with the supporting signal/communications company and the executive officer, ensures that the critical network maintenance is performed and that parts are available as needed for communications systems to remain operational.

COMMUNICATIONS SECURITY MAINTENANCE

H-40. COMSEC equipment is evacuated through normal maintenance channels to the brigade support battalion or the brigade signal company—or through the local electronic key management system for the Marine Corps, if appropriate. Managing COMSEC measures, including the operation of the information system security office of the communications support elements.

H-41. Items procured under the National Security Agency Commercial COMSEC Evaluation Program are fielded with a limited vendor warranty. All COMSEC equipment with a vendor warranty is maintained and serviced by the original equipment manufacturer for sustainment support. Once the vendor warranty expires, all sustainment repairs are transitioned to Tobyhanna Army Depot for organic support in the Army. The Marine Corps releases a naval message directing the equipment to a supporting establishment for future repairs.

ISOLATION OF FAULTS AND CONTROLLED EXCHANGE

H-42. The communications equipment operator/maintainer performs field maintenance on the communications equipment. This includes performing preventive maintenance and evaluating the cause of equipment failure through troubleshooting and the use of built-in test equipment. The operator/maintainer is also responsible for the minor repair (check, adjust, tighten) and removal and replacement of unserviceable line replaceable units/line replaceable modules. The unserviceable line replaceable unit/line replaceable module is then evacuated using established procedures for repair or replacement. This is accomplished through the supporting maintenance element.

H-43. Controlled exchange refers to the removal of serviceable parts from an item of non-mission-capable equipment for installation on another piece of equipment that can quickly or easily be rendered mission-capable. The TTSB/MAGTF G-6 SOP may give battalion commanders the authority to direct control exchanges as long as controlled substitutions are conducted according to AR 750-1 or annex K to the Marine expeditionary force OPORD. Controlled exchange is done under the direction of the commander based on the recommendation of the S-6 and/or the signal/communications company commander. Controlled exchange may be done at any echelon under these guidelines.

This page intentionally left blank.

Glossary

The glossary lists acronyms and terms with Army or joint definitions. Where Army and joint definitions differ, (Army) precedes the definition. Terms for which ATP 3-37.10/MCRP 3-40D.13 is the proponent are marked with an asterisk (*). The proponent publication for other terms is listed in parentheses after the definition.

SECTION I – ACRONYMS AND ABBREVIATIONS

ADRP	Army doctrine reference publication
AFCEC	Air Force Civil Engineer Center
AFCS	Army Facilities Components System
AFPAM	Air Force pamphlet
AFMAN	Air Force manual
AFTTP	Air Force tactics, techniques, and procedures
AO	area of operations
AOR	area of responsibility
AR	Army regulation
AT	antiterrorism
attn	attention
ATP	Army techniques publication
BCOC	base cluster operations center
BDOC	base defense operations center
BOC	base operations center
BOS-I	base operating support-integrator
BCT	brigade combat team
CAAF	contractors authorized to accompany the force
CBRN	chemical, biological, radiological, and nuclear
CCDR	combatant commander
CCIR	commander's critical information requirement
CCMD	combatant command
CID	criminal investigation division
CMU	concrete masonry unit
COA	course of action
COMSEC	communications security
COP	common operational picture
COR	contracting officer representative
CSS	combat service support
DA	Department of the Army
DC	District of Columbia
DD	Department of Defense

Glossary

DOD	Department of Defense
DODD	Department of Defense directive
DODI	Department of Defense instruction
DRSN	Defense Red Switched Network
DSN	Defense Switched Network
EBS	environmental baseline survey
ECP	entry control point
EO	executive order
EP	engineer pamphlet
ERDC	Engineer Research and Development Center
ESB	expeditionary signal battalion
ESOH	environmental safety and occupational health
FC	facilities criteria
FFCC	force fires coordination cell
FM	field manual
G-1	assistant chief of staff, personnel
G-2	assistant chief of staff, intelligence
G-3	assistant chief of staff, operations
G-4	assistant chief of staff, logistics
G-6	assistant chief of staff for communications
G-8	assistant chief of staff, financial management
G-9	assistant chief of staff, civil affairs operations
GTA	graphic training aid
HAZMAT	hazardous materials
HN	host nation
HW	hazardous waste
IED	improvised explosive device
IMCOM	United States Army Installation Management Command
IPB	intelligence preparation of the battlefield (Army)/intelligence preparation of the battlespace (Marine Corps)
J-6	communications system directorate of a joint staff
JCMS	joint construction management system
JFC	joint force commander
JOPP	joint operation planning process
JP	joint publication
KOCOA	key terrain, observation and fields of fire, cover and concealment, obstacles, and avenues of approach (Marine Corps)
LAN	local area network
LOC	line of communications
LOGCAP	sustainment/logistics civil augmentation program (Army)
MAGTF	Marine air-ground task force
MCPP	Marine Corps planning process

Glossary

MCRP	Marine Corps reference publication
MCWP	Marine Corps warfighting publication
MDMP	military decision-making process
MEB	maneuver enhancement brigade
METT-T	mission, enemy, terrain and weather, troops and support available–time available (Marine Corps)
METT-TC	mission, enemy, terrain and weather, troops and support available–time available and civil considerations (Army)
MILCON	military construction
MIL-STD	military standard
MO	Missouri
MSCoE	Maneuver Support Center of Excellence
MSF	mobile security force
MWR	morale, welfare, and recreation
NATO	North Atlantic Treaty Organization
NAVFAC	Naval Facilities Engineering Command
NETOPS	network operations
NIPRNET	nonsecure Internet protocol router network
No.	number
NTTP	Navy tactics, techniques, and procedures
NTRP	Navy tactical reference publication
O&M	operation and maintenance
OAKOC	observation and fields of fire, avenues of approach, key terrain, obstacles, and cover and concealment
OEHSA	occupational and environmental health site assessment
OP	observation post
OPLAN	operation plan
OPORD	operation order
PVNTMED	preventive medicine
PWS	performance work statement
QOL	quality of life
RCT	regimental combat team
RFI	request for information
REDi	reachback engineer data integration
RO	reverse osmosis
ROE	rules of engagement
S-1	battalion or brigade manpower and personnel staff officer (Marine Corps battalion or regiment)
S-2	battalion or brigade intelligence staff officer (Army, Marine Corps battalion or regiment)
S-3	battalion or brigade operations staff officer (Army; Marine Corps battalion or regiment)

Glossary

S-4	battalion or brigade sustainment/logistics staff officer (Army; Marine Corps battalion or regiment)
S-6	battalion or brigade communications staff officer (Army; Marine Corps battalion or regiment)
S-9	battalion or brigade civil affairs operations staff officer
SA	situational awareness
SEA	Southeast Asia
SIPRNET	secret Internet protocol router network
SOP	standard operating procedure
SWA	Southwest Asia
SU	situational understanding (Army)
TCF	tactical combat force
TEMPER	tent extendible modular personnel
TM	technical manual
TTP	tactics, techniques, and procedures
TTSB	theater tactical signal brigade
UAS	unmanned aircraft system
UFC	Unified Facilities Criteria
UROC	United States Army Corps of Engineers Reachback Operations Center
U.S.	United States
USACE	United States Army Corps of Engineers
USAMEDCOM	United States Army Medical Command
USC	United States Code
VoIP	voice over Internet protocol
WAN	wide-area network

SECTION II – TERMS

*base camp
    An evolving military facility that supports the military operations of a deployed unit and provides the necessary support and services for sustained operations.

References

REQUIRED PUBLICATIONS
These documents must be available to the intended users of this publication.
> ADRP 1-02. *Terms and Military Symbols*. 16 November 2016.
> DOD Dictionary of Military and Associated Terms. 15 October 2016.
> MCRP 1-10.2. *Marine Corps Supplement to the Department of Defense Dictionary of Military and Associated Terms*. 16 November 2011.

RELATED PUBLICATIONS
These documents contain relevant supplemental information.

ARMY
Most Army publications are available online at <https://armypubs.army.mil>.
> ADP 5-0. *The Operations Process*. 17 May 2012.
> ADRP 3-0. *Unified Land Operations*. 11 November 2016.
> ADRP 3-37. *Protection*. 31 August 2012.
> ADRP 3-90. *Offense and Defense*. 31 August 2012.
> ADRP 4-0. *Sustainment*. 31 July 2012.
> ADRP 5-0. *The Operations Process*. 17 May 2012.
> ADRP 6-0. *Mission Command*. 17 May 2012.
> ADRP 7-0. *Training Units and Developing Leaders*. 23 August 2012.
> AR 25-30. *Army Publishing Program*. 3 June 2015.
> AR 200-1. *Environmental Protection and Enhancement*. 13 December 2007.
> AR 210-20. *Real Property Master Planning for Army Installations*. 16 May 2005.
> AR 420-1. *Army Facilities Management*. 24 August 2012.
> AR 735-5. *Property Accountability Policies*. 10 May 2013.
> AR 750-1. *Army Materiel Maintenance Policy*. 12 September 2013.
> ATP 3-34.22. *Engineer Operations–Brigade Combat Team and Below*. 5 December 2014.
> ATP 3-34.23. *Engineer Operations–Echelons Above Brigade Combat Team*. 10 June 2015.
> ATP 3-34.80. *Geospatial Engineering*. 23 June 2014.
> ATP 3-37.2. *Antiterrorism*. 3 June 2014.
> ATP 3-39.10. *Police Operations*. 26 January 2015.
> ATP 3-39.30. *Security and Mobility Support*. 30 October 2014.
> ATP 3-39.32. *Physical Security*. 30 April 2014.
> ATP 3-39.34. *Military Working Dogs*. 30 January 2015.
> ATP 3-39.35. *Protective Services*. 31 May 2013.
> ATP 4-02.1. *Army Medical Logistics*. 29 October 2015.
> ATP 4-25.12. *Unit Field Sanitation Team*. 30 April 2014.
> ATP 4-32. *Explosive Ordnance Disposal (EOD) Operations*. 30 September 2013.
> ATP 4-35.1. *Ammunition and Explosive Handler Safety Techniques*. 8 November 2016.
> ATP 4-42. *General Supply and Field Services Operations*. 14 July 2014.

References

ATP 4-45. *Force Provider Operations.* 24 November 2014.
ATP 4-92. *Contracting Support to Unified Land Operations.* 15 October 2014.
ATP 5-19. *Risk Management.* 14 April 2014.
DA Pamphlet 385-30. *Risk Management.* 2 December 2014.
DA Pamphlet 385-64. *Ammunition and Explosives Safety Standards.* 24 May 2011.
EP 500-1-2. *Emergency Employment of Army and Other Resources - Field Force Engineering - United States Army Corps of Engineers Support to Full Spectrum Operations.* 1 August 2010. <http://www.publications.usace.army.mil/Portals/76/Publications/EngineerPamphlets/EP_500-1-2.pdf?ver=2013-08-22-090239-343>, accessed 29 November 2016.
EP 1105-3-1. *Base Camp Development in the Theater of Operations.* 19 January 2009. Available at <http://www.publications.usace.army.mil/Portals/76/Publications/EngineerPamphlets/EP_1105-3-1.pdf>, accessed 9 September 2016.
FM 1-0. *Human Resources Support.* 1 April 2014.
FM 1-04. *Legal Support to the Operational Army.* 18 March 2013.
FM 1-06. *Financial Management Operations.* 15 April 2014.
FM 2-0. *Intelligence Operations.* 15 April 2014.
FM 3-01. *U.S. Army Air and Missile Defense Operations.* 2 November 2015.
FM 3-09. *Field Artillery Operations and Fire Support.* 4 April 2014.
FM 3-11. *Multiservice Doctrine for Chemical, Biological, Radiological, and Nuclear Operations,* 1 July 2011.
FM 3-34. *Engineer Operations.* 2 April 2014.
FM 3-39. *Military Police Operations.* 26 August 2013.
FM 3-57. *Civil Affairs Operations.* 31 October 2011.
FM 3-90-1. *Offense and Defense, Volume 1.* 22 March 2013.
FM 3-90-2. *Reconnaissance, Security, and Tactical Enabling Tasks, Volume 2.* 22 March 2013.
FM 4-02. *Army Health System.* 26 August 2013.
FM 4-40. *Quartermaster Operations.* 22 October 2013.
FM 6-0. *Commander and Staff Organization and Operations.* 5 May 2014.
FM 6-02. *Signal Support Operations.* 22 January 2014.
FM 27-10. *The Law of Land Warfare.* 18 July 1956.
GTA 05-08-016. *The Environment and Redeployment: How to Transition a Base Camp.* 1 March 2012.
GTA 05-08-018. *Dust Suppression Techniques.* 7 April 2016.
GTA 90-01-011. *Joint Forward Operations Base (JFOB) Protection Handbook, Sixth Edition.* 1 October 2011.
GTA 90-01-018. *Joint Entry Control Point & Escalation of Force Procedures (JEEP) Handbook.* 1 December 2009.
GTA 90-01-034. *Small-Base Entry Control Point Guide: A Practical Guide for the Small-Base Leader.* 1 April 2012.
TC 4-02.3. *Field Hygiene and Sanitation.* 6 May 2015.
TM 3-34.30. *Firefighting.* 23 April 2015.
TM 3-34.45. *Engineer Prime Power Operations.* 13 August 2013.
TM 3-34.48-1. *Theater of Operations: Roads, Airfields, and Heliports—Road Design.* 29 February 2016.
TM 3-34.48-2. *Theater of Operations: Roads, Airfields, and Heliports—Airfield and Heliport Design.* 29 February 2016.

TM 3-34.55. *Construction Surveying.* 12 August 2012.

TM 3-34.63. *Paving and Surfacing Operations.* 13 August 2013.

TM 5-304. *Army Facilities Components System User Guide.* 1 October 1990.

TM 5-610. *Preventive Maintenance for Facilities Engineering, Buildings and Structures.* 1 November 1979.

TM 10-4320-303-13. *Tactical Water Distribution Equipment System (TWDS) Set 10-Mile Segment.* 30 June 1993.

TM 38-410. *Storage and Handling of Hazardous Materials.* 13 January 1999.

DEPARTMENT OF DEFENSE PUBLICATIONS

MIL-STD-2525D. *Joint Military Symbology.* 10 June 2014.
<http://www.dtic.mil/doctrine/doctrine/other/ms_2525d.pdf>, accessed on 10 August 2016.

DODD 3000.10. *Contingency Basing Outside the United States.* 10 January 2013.
<http://www.dtic.mil/whs/directives/corres/pdf/300010p.pdf>, accessed on 10 August 2016.

DODD 4270.5. *Military Construction.* 16 June 2005.
<https://www.wbdg.org/ccb/ARMYCOE/COEECB/ARCHIVES/ecb_2005_8.pdf>, accessed on 10 August 2016.

DODD 4705.1E. *Management of Land-Based Water Resources in Support of Contingency Operations* 3 June 2015. <http://www.dtic.mil/whs/directives/corres/pdf/470501p.pdf>, accessed on 10 August 2016.

DODI 3020.50. *Private Security Contractors (PSCs) Operating in Contingency Operations.* 22 July 2009 <http://www.dtic.mil/whs/directives/corres/pdf/302050p.pdf>, accessed on 10 August 2016.

DODI 4165.70. *Real Property Management.* 6 April 2005.
<http://www.dtic.mil/whs/directives/corres/pdf/416570p.pdf>, accessed on 10 August 2016.

JOINT

Most joint publications are available online at <www.dtic.mil/doctrine/new_pubs/jointpub.htm>.

JP 1-04. *Legal Support to Military Operations.* 2 August 2016.

JP 3-0. *Joint Operations.* 11 August 2011.

JP 3-07.2. *Antiterrorism.* 14 March 2014.

JP 3-10. *Joint Security Operations in Theater.* 13 November 2014.

JP 3-15. *Barriers, Obstacles, and Mine Warfare for Joint Operations.* 17 June 2011.

JP 3-34. *Joint Engineer Operations.* 6 January 2016.

JP 3-41. *Chemical, Biological, Radiological, and Nuclear, Consequence Management.* 21 June 2012.

JP 4-0. *Joint Logistics.* 16 October 2013.

JP 4-10. *Operational Contract Support.* 16 July 2014.

JP 5-0. *Joint Operation Planning.* 11 August 2011.

MARINE CORPS PUBLICATIONS

Most Marine Corps publications are available online at <http://www.marines.mil/News/Publications/ELECTRONICLIBRARY.aspx>.

MCDP 1-0. *Marine Corps Operations.* 9 August 2011.

MCDP 6. *Command and Control.* 4 October 1996.

MCRP 3-40B.3. *Contingency Contracting.* 12 February 2009.

MCRP 3-40D.2. *Planning and Design of Roads, Airfields, and Heliports in the Theater of Operations—Airfield and Heliport Design.* 29 September 1994.

MCRP 8-10B. *How to Conduct Training.* 10 August 2015.

References

MCTP 3-40H. *MAGTF Supply Operations.* 29 February 1996.
MCTP 8-10A. *Unit Training Management Guide.* 25 November 1996.
MCTP 8-10B. *How to Conduct Training.* 10 August 2015.
MCTP 10-10F. *Military Police Operations.* 9 September 2010.
MCWP 3-34. *Engineering Operations.* 14 February 2000.
MCWP 5-10. *Marine Corps Planning Process.* 24 August 2010.

Multi-Service Publications

ATP 3-09.32/MCRP 3-31.6/NTTP 3-09.2/AFTTP (I) 3-2.6. *JFIRE Multi-Service Tactics, Techniques, and Procedures for the Joint Application of Firepower.* 21 January 2016.
ATP 3-11.37/MCRP 10-10.F7/NTTP 3-11.29/AFTTP 3-2.44. *Multi-Service Tactics, Techniques, and Procedures for Chemical Biological, Radiological, and Nuclear Reconnaissance and Surveillance.* 25 March 2013.
ATP 3-17.2/MCRP 3-20B.1/NTTP 3-02.18/AFTTP 3-2.68. *Multi-Service Tactics, Techniques, and Procedures for Airfield Opening.* 18 June 2015.
ATP 3-34.5/MCRP 3-40B.2. *Environmental Considerations.* 10 August 2015.
ATP 3-34.40/MCTP 3-40D. *General Engineering.* 25 February 2015.
ATP 3-34.81/MCRP 3-34.3. *Engineer Reconnaissance.* 1 March 2016.
ATP 3-37.34/MCTP 3-34C. *Survivability Operations.* 28 June 2013.
ATP 3-90.4/MCWP 3-34A. *Combined Arms Mobility.* 8 March 2016.
ATP 3-90.8/MCTP 3-34B. *Combined Arms Countermobility.* 17 September 2014.
ATP 4-10/MCRP 3-40B.6/NTTP 4-09.1/AFMAN 10-409-O. *Multi-Service Tactics Techniques and Procedures for Contract Support.* 18 February 2016.
ATP 4-44/MCRP 3-40D.14. *Water Support Operations.* 2 October 2015.
NTRP 4-04.2.3/TM 3-34.41/AFPAM 32-1000/MCRP 3-40D.12. *Construction Estimating.* 1 December 2010.
NTRP 4-04.2.5/TM 3-34.42/AFPAM 32-1020/MCRP 3-40D.6. *Construction Project Management.* 1 December 2012.
NTTP 3-10.1M/MCWP 4-11.5. *Seabee Operations in the Marine Air-Ground Task Force (MAGTF).* March 2015.
TM 3-11.42/MCTP 10-10G/NTTP 3-11.36/AFTTP 3-2.83. *Multi-Service Tactics, Techniques, and Procedures for Installation Emergency Management.* 23 June 2014.
TM 3-34.44/ MCRP 3-40D.4. *Concrete and Masonry.* 23 July 2012.
TM 3-34.46/MCRP 3-40D.11. *Theater of Operations Electrical Systems.* 3 May 2013.
TM 3-34.47/MCRP 3-17C. *Carpentry.* 20 September 2013.
TM 3-34.49/NAVFAC P-1065/AFMAN 32-1072. *Multiservice Procedures for Well-Drilling Operations.* 1 December 2008.
TM 3-34.56/MCIP 3-40G.2i. *Waste Management for Deployed Forces.* 19 July 2013.
TM 3-34.62/MCRP 3-40D.9. *Earthmoving Operations.* 29 June 2012.
TM 3-34.64/MCRP 3-40D.7. *Military Soils Engineering.* 25 September 2012.
TM 3-34.70/MCRP 3-40D.5. *Plumbing, Pipe Fitting, and Sewerage.* 23 July 2012.
TM 3-34.85/MCRP 3-34.1. *Engineer Field Data.* 17 October 2013.

Other Publications

AFPAM 10-219 Volume 4. *Airfield Damage Repair Operations.* 28 May 2008. <http://static.e-publishing.af.mil/production/1/af_a4/publication/afpam10-219v4/afpam10-219v4.pdf>, accessed 29 November 2016.

AFPAM 10-219 Volume 7. *Expedient Methods.* 9 June 2008.
< http://static.e-publishing.af.mil/production/1/af_a4/publication/afpam10-219v7/afpam10-219v7.pdf>, accessed on 29 November 2016.

Chairman of the Joint Chiefs of Staff Instruction (CJCSI) 3121.01B. *Standing Rules of Engagement/Standing Rules for the Use of Force for U.S. Forces.* 13 June 2005.
<http://www.loc.gov/rr/frd/Military_Law/pdf/OLH_2015_Ch5.pdf >, accessed on 9 August 2016.

EO 13112. *Invasive Species.* 3 February 1999.
<https://www.gpo.gov/fdsys/pkg/FR-1999-02-08/pdf/99-3184.pdf>, accessed on 10 August 2016.

FC 4-722-01F. *Air Force Dining Facilities.* 2 July 2007.
<https://www.wbdg.org/ccb/DOD/UFC/fc_4_722_01f.pdf>, accessed on 10 August 2016.

Federal Standard 376B. Preferred Metric Units for General Use for the Federal Government. 27 January 1993. <http://www.nist.gov/pml/wmd/metric/upload/fs376-b.pdf>, accessed on 10 August 2016.

OSHA *Field Safety and Health Manual.* 23 May 2011.
<https://www.osha.gov/OshDoc/Directive_pdf/ADM_04-00-001.pdf>, accessed on 10 August 2016.

Posse Comitatus Act. <*www.dtic.mil/dtic/tr/fulltext/u2/a494995.pdf*>, accessed on 10 August 2016.

10 USC. *Armed Forces.* Edition July 2011.
<https://www.gpo.gov/fdsys/pkg/CPRT-112HPRT67344/pdf/CPRT-112HPRT67344.pdf>, accessed on 10 August 2016.

10 USC 2808. *Armed Forces.* <http://uscode.house.gov/browse/&edition=prelim>, accessed 30 December 2016.

UFC 1-200-01. *General Building Requirements.* 20 June 2016.
<http://www.wbdg.org/ccb/DOD/UFC/ufc_1_200_01.pdf>, accessed on 10 August 2016.

UFC 1-201-01. *Non-Permanent DOD Facilities in Support of Military Operations.* 1 January 2013.
<https://www.wbdg.org/ccb/DOD/UFC/ufc_1_201_01.pdf>, accessed on 10 August 2016.

UFC 1-201-02. *Unified Facilities Criteria Assessment of Existing Facilities for use in Military Operations.* 1 March 2013. <https://www.wbdg.org/ccb/DOD/UFC/ufc_1_201_02.pdf>, accessed on 10 August 2016.

UFC 2-000-05N. *Facility Planning for Navy and Marine Corps Shore Installations.* 31 January 2005.
< https://www.wbdg.org/ccb/browse_docex.php?d=7226>, accessed on 26 May 2016.

UFC 2-100-01. *Installation Master Planning.* 15 May 2012.
<http://wbdg.org/ccb/DOD/UFC/ufc_2_100_01.pdf>, accessed on 10 August 2016.

UFC 3-190-06. *Protective Coatings and Paints.* 16 January 2004.
<https://www.wbdg.org/ccb/DOD/UFC/ufc_3_190_06.pdf>, accessed on 10 August 2016.

UFC 3-230-01. *Water Storage, Distribution, and Transmission.* 1 November 2012.
<https://www.wbdg.org/ccb/DOD/UFC/ufc_3_230_01.pdf>, accessed on 10 August 2016.

UFC 3-230-03. *Water Treatment.* 1 November 2012.
<https://www.wbdg.org/ccb/DOD/UFC/ufc_3_230_03.pdf>, accessed on 10 August 2016.

UFC 3-260-01. *Airfield and Heliport Planning and Design.* 17 November 2008.
<https://www.wbdg.org/ccb/DOD/UFC/ufc_3_260_01.pdf>, accessed on 10 August 2016.

UFC 3-260-17. *Dust Control for Roads, Airfields and Adjacent Areas.* 16 January 2004.
<https://www.wbdg.org/ccb/DOD/UFC/ufc_3_260_17.pdf>, accessed on 10 August 2016.

UFC 3-600-01. *Fire Protection Engineering for Facilities.* 8 August 2016.
<https://www.wbdg.org/ccb/DOD/UFC/ufc_3_600_01.pdf>, accessed on 10 August 2016.

UFC 4-010-01. *DOD Minimum Antiterrorism Standards for Buildings.* 9 February 2012.

References

<https://www.wbdg.org/ccb/DOD/UFC/ufc_4_010_01.pdf>, accessed on 10 August 2016.

UFC 4-010-02. *DOD Minimum Antiterrorism Standoff Distances for Buildings.* 9 February 2009.
< http://www.wbdg.org/ffc/dod/unified-facilities-criteria-ufc/ufc-4-010-02, accessed on 29 November 2016.

UFC 4-020-01. *DOD Security Engineering Facilities Planning Manual.* 11 September 2008.
<https://www.wbdg.org/ccb/DOD/UFC/ufc_4_020_01.pdf>, accessed on 10 August 2016.

UFC 4-021-02. *Electronic Security Systems.* 1 October 2013.
<https://www.wbdg.org/ccb/DOD/UFC/ufc_4_021_02.pdf>, accessed on 10 August 2016.

UFC 4-022-01. *Security Engineering: Entry Control Facilities/Access Control Points.* 25 May 2005.
<https://www.wbdg.org/ccb/DOD/UFC/ufc_4_022_01.pdf>, accessed on 10 August 2016.

UFC 4-022-02. *Selection and Application of Vehicle Barriers.* 8 June 2009.
<https://www.wbdg.org/ccb/DOD/UFC/ufc_4_022_02.pdf>, accessed on 10 August 2016.

UFC 4-022-03. *Security Fence and Gates.* 1 October 2013.
<https://www.wbdg.org/ccb/DOD/UFC/ufc_4_022_03.pdf>, accessed on 10 August 2016.

UFC 4-451-10N. *Design: Hazardous Waste Storage.* 16 January 2004.
<https://www.wbdg.org/ccb/DOD/UFC/ufc_4_451_10n.pdf>, accessed on 10 August 2016.

UFC 4-510-01. *Design: Military Medical Facilities.* 1 May 2016.
<http://www.wbdg.org/ccb/DOD/UFC/ufc_4_510_01.pdf>, accessed on 10 August 2016.

Uniform Code of Military Justice. <http://www.ucmj.us/>, accessed on 30 December 2016.

Whole Building Design Guide (UFC Index), <http://www.wbdg.org>, accessed on 10 August 2016.

PRESCRIBED FORMS

None.

REFERENCED FORMS

Unless otherwise indicated, DA forms are available on the Army Publishing Directorate Web site at <https://armypubs.army.mil>. DD forms are available on the Office of the Secretary of Defense Web site at <http://:www.dtic.mil/whs/directives/forms/index.htm>.

DA Form 2028. *Recommended Changes to Publications and Blank Forms.*

DD Form 1354. *Transfer and Acceptance of DOD Real Property.*

DD Form 1391. *FY __ Military Construction Project Data.*

DD Form 2993. *Environmental Baseline Survey (EBS) Checklist.*

DD Form 2994. *Environmental Baseline Survey (EBS) Report.*

DD Form 2995. *Environmental Site Closure Survey.*

WEB SITES

Air Force Civil Engineer Center Web site, <http://www.afcec.af.mil/>, accessed on 10 August 2016.

Army Geospatial Center, <http://www.agc.army.mil/>, accessed on 10 August 2016.

Army Knowledge Online, Doctrine and Training Publications Web site, <https://armypubs.us.army.mil/doctrine/index.html>, accessed on 10 August 2016.

Army Publishing Directorate, Army Publishing Updates Web site, <http://www.apd.army.mil/AdminPubs/new_subscribe.asp>, accessed on 10 August 2016.

Construction Criteria Base System maintained by the National Institute of Building Sciences at <http://www.nibs.org/index.php/ccb>, accessed 10 August 2016.

Health Facilities Planning Agency, Office of the Surgeon General, <http://www.armyhealthfacilities.amedd.army.mil/>, accessed 10 August 2016.

International Civil Aviation Organization, <http://www.icao.int/Pages/default.aspx>, accessed 10 August 2016.
Naval Facilities Engineering Command (NAVFAC), <http://www.navfac.navy.mil>, accessed on 10 August 2016.
NAVFAC Reachback Information Portal, <https://hub.navfac.navy.mil>, accessed on 10 August 2016.
Title 10 USC, <http://uscode.house.gov/browse/prelim@title10&edition=prelim>, accessed on 10 August 2016.
USACE Reachback Engineer Data Integration, <https://redi.usace.army.mil>, accessed on 10 August 2016.
U.S. Army Corps of Engineers Reachback Operations Center (UROC), <https://uroc.usace.army.mil>, accessed on 10 August 2016.
USACE Engineer Research and Development Center (ERDC) Force Protection Portal, <https://erdc-fp.usace.army.mil>, accessed on 10 August 2016.
USACE Official Publications Web site, <http://www.publications.usace.army.mil/USACEPublications/EngineerPamphlets.aspx?udt_43545_param_page=3>, accessed on 10 August 2016.
USACE Protective Design Center Web site, <https://pdc.usace.army.mil/library>, accessed on 10 August 2016.
USACE technical information <http://www.hnd.usace.army.mil/techinfo>, accessed on 10 August 2016.

RECOMMENDED READINGS

These documents contain relevant supplemental information.
AFH 10-222. *Guide to Bare Base Assets, Volume 2*. 6 February 2012.
AFH 10-222. *Contingency Water System Installation and Operation, Volume 11*. 19 May 2011.
AFPAM 10-219 Volume 5. *Bare Base Conceptual Planning*. 30 March 2012.
   <http://wbdg.org/ccb/AF/AFP/afpam10_219_v5.pdf>, accessed on 10 August 2016.
DA Pamphlet 420-1-2. *Army Military Construction and Nonappropriated-Funded Construction Program Development and Execution*. 2 April 2009.
FM 3-81. *Maneuver Enhancement Brigade*. 21 April 2014.
MCTP 3-34. *Engineering Operations*. 14 February 2000.
Title 50 USC 1621. *Declaration of National Emergency by President*. 3 January 2012.
   <https://www.gpo.gov/fdsys/pkg/USCODE-2011-title50/pdf/USCODE-2011-title50-chap15-sec401.pdf>, accessed on 10 August 2016.
UFC 3-230-02. *Operations and Maintenance: Water Supply Systems*. 10 July 2001.
   <https://www.wbdg.org/ccb/DOD/UFC/ufc_3_230_02.pdf>, accessed on 10 August 2016.
UFC 3-501-01. *Electrical Engineering*. 6 October 2015.
   <https://www.wbdg.org/ccb/DOD/UFC/ufc_3_501_01.pdf>, accessed on 10 August 2016.
UFC 3-510-01. *Foreign Voltages and Frequencies Guide*. 1 March 2005.
   <https://www.wbdg.org/ccb/DOD/UFC/ufc_3_510_01.pdf>, accessed on 10 August 2016.
UFC 3-520-01. *Interior Electrical Systems*. 6 October 2015.
   <https://www.wbdg.org/ccb/DOD/UFC/ufc_3_520_01.pdf>, accessed on 10 August 2016.
UFC 3-520-05. *Stationary Battery Areas*. 1 May 2015.
   <https://www.wbdg.org/ccb/DOD/UFC/ufc_3_520_05.pdf>, accessed on 10 August 2016.
UFC 3-530-01. *Interior and Exterior Lighting Systems and Controls*. 1 April 2015.
   <https://www.wbdg.org/ccb/DOD/UFC/ufc_3_530_01.pdf>, accessed on 10 August 2016.

References

UFC 3-535-01. *Visual Air Navigation Facilities*. 17 December 2005.
<https://www.wbdg.org/ccb/DOD/UFC/ufc_3_535_01.pdf>, accessed on 10 August 2016.

UFC 3-540-01. *Engine-Driven Generator Systems for Backup Power Applications*. 1 August 2014.
<https://www.wbdg.org/ccb/DOD/UFC/ufc_3_540_01.pdf>, accessed on 10 August 2016.

UFC 3-550-01. *Exterior Electrical Power Distribution*. 1 September 2016.
<https://www.wbdg.org/ccb/DOD/UFC/ufc_3_550_01.pdf>, accessed on 10 August 2016.

UFC 3-555-01N. 400 *Hertz Medium Voltage Conversion/Distribution and Low Voltage Utilization Systems*. 16 January 2004. <https://www.wbdg.org/ccb/DOD/UFC/ufc_3_555_01n.pdf>, accessed on 10 August 2016.

UFC 3-560-01. *Electrical Safety, O&M*. 6 December 2006.
<https://www.wbdg.org/ccb/DOD/UFC/ufc_3_560_01.pdf>, accessed on 10 August 2016.

UFC 3-570-02A. *Cathodic Protection*. 1 March 2005.
<https://www.wbdg.org/ccb/DOD/UFC/ufc_3_570_02a.pdf>, accessed on 10 August 2016.

UFC 3-570-02N. *Electrical Engineering Cathodic Protection*. 16 January 2004.
<https://www.wbdg.org/ccb/DOD/UFC/ufc_3_570_02n.pdf>, accessed on 10 August 2016.

UFC 3-570-06. *O&M: Cathodic Protection Systems*. 31 January 2003.
<https://www.wbdg.org/ccb/DOD/UFC/ufc_3_570_06.pdf>, accessed on 10 August 2016.

UFC 3-575-01. *Lightning and Static Electricity Protection Systems*. 1 July 2012.
<https://www.wbdg.org/ccb/DOD/UFC/ufc_3_575_01.pdf>, accessed on 10 August 2016.

UFC 3-580-01. *Telecommunications Interior Infrastructure Planning and Design*. 1 June 2016.
<https://www.wbdg.org/ccb/DOD/UFC/ufc_3_580_01.pdf>, accessed on 10 August 2016.

UFC 3-730-01. *Programming Cost Estimates for Military Construction*. 6 June 2011.
<http://www.wbdg.org/ffc/dod/unified-facilities-criteria-ufc/ufc-3-730-01>, accessed on 29 November 2016.

Index

Entries are by paragraph number.

A
antiterrorism (AT), F-16

B
base camp
  appendix, C-1
  basing strategy, 2-9, 2-60
  classification system, 1-17
  closure, 1-43, 2-57
  construction, 1-23
  development planning process, 2-13, B-4
  life cycle, 1-37
  management center, 1-46
  planning (See also basing strategy), 2-9, 2-25, 2-60
  planning factors, E-1, E-2, E-9
  roles and responsibilities for, 1-70
  scheme of, 2-60
  services, 1-68
  site layout, 3-26, D-1
  sizes of, 1-21, E-9
  standards, 1-22, 1-23
  working groups, 1-46, 1-106
base cluster, 1-10
base cluster operations center (BCOC), 1-46
base operations center (BOC), 1-46
billeting, D-7, F-29

C
communications support (See signal support), H-1, H-11
construction
  base camps, 1-23
  standards, 1-23

contracted support, H-10

D
design
  electrical power and distribution systems, F-61
  water production and distribution, F-47
design process, F-1
dining facilities (DFACs), F-31
drainage, F-26

E
emergency management, 4-44
engineer staff officer, B-32
environmental considerations, B-16, D-9

F
force protection (FP) (See antiterrorism [AT]), F-16

I
intelligence preparation of the battlefield/battlespace (IPB), B-7, B-22

L
land use categories, D-7
land use planning (See also planning), 2-47, D-1, D-4, D-7

M
master planning, 2-27, 2-29
medical treatment facilities, E-43
mission analysis, B-4, B-6
mission variables, B-8

mission, enemy, terrain and weather, troops and support available, time available, civilian considerations (METT-TC) (see mission variables), B-8
motor pools, F-33

O
operational art, 1-13
operational environment, 1-8

P
planning (See also site selection)
  strategic (See also basing strategy), 2-9, E-7

R
real estate, B-10
risk management (RM), B-35

S
safety, F-19
signal support, H-1, H-11
site selection (See also planning), B-8

T
terrain analysis, B-13
toilet and shower facilities, F-30
traffic control, F-37

U
United States Army Corps of Engineers (USACE), 1-70, B-26, F-66

This page intentionally left blank.

ATP 3-37.10
27 January 2017

By Order of the Secretary of the Army:

MARK A. MILLEY
General, United States Army
Chief of Staff

Official:

GERALD B. O'KEEFE
Administrative Assistant to the
Secretary of the Army
1702406

DISTRIBUTION:
Active Army, Army National Guard, and United States Army Reserve: Distributed in electronic media only (EMO).

PCN: 144 000201 00   PIN: 103345–000

www.ingramcontent.com/pod-product-compliance
Lightning Source LLC
Chambersburg PA
CBHW050057230526
45470CB00004B/1568